THE COMPLETE BEE HANDBOOK

THE COMPLETE BEE HANDBOOK

HISTORY, RECIPES, BEEKEEPING BASICS, AND MORE

Dr. Dewey M. Caron

ROCKRIDGE PRESS

For general information on our other products and services or to obtain technical support, please contact our Customer Care Department within the United States at (866) 744-2665, or outside the United States at (510) 253-0500.

Rockridge Press publishes its books in a variety of electronic and print formats. Some content that appears in print may not be available in electronic books, and vice versa.

Interior and Cover Designer: Michael Patti
Art Producer: Tom Hood
Editor: Sam Eichner

Illustrations © Jyotirmayee Patra 2020. Photographs Temmuz can Arsiray/Shutterstock, pp. v, 49, 71, 182; Anton Starikov/Alamy p. 6; Life on White/Alamy, p. 6; The Natural History Museum/Science Source, p. 7; blickwinkel/ Alamy, p. 13; Matthew Cole/Shutterstock, p. 20; Geoff Smith/Alamy, p. 24; dpa picture alliance/ Alamy, p. 36; grafxart8888/iStock, p. 43; PEIYU/ Shutterstock, p. 60; Anna Chaplygina/iStock, p. 63; Carolyn Franks / Alamy, p. 64; Tomentosaplaga/Alamy, p. 83; Shutterstock, pp. 87, 96; JONATHAN WILKINS/Science Source, p. 89; Craig Steven Thrasher/Alamy, p. 90; Janet Sheppardson/Alamy, p. 103; DLugowski/Alamy, p. 119; OcsanaDen/iStock, p. 124; Pluto/Alamy, pp. 128, 129; Olesia Shadrina/iStock, p. 133; viennetta/iStock, p. 140; Karaidel/iStock, p. 142; Pinkybird/iStock, p. 148; Galiyah Assan/ Shutterstock, p. 150; bhofack2/iStock, p. 156; Alla Rudenko/Alamy, p. 160; Mayiorica/Shutterstock, p. 164; Rawpixel.com/iStock, p. 168.

ISBN: Print 978-1-64611-987-5 |
eBook 978-1-64611-988-2

R0

I dedicate this book
to keepers of bees,
in fact or wannabee.
Beekeeping is a journey.
May your journey
be fulfilling.

Contents

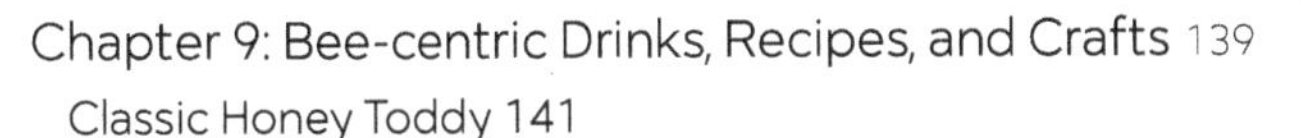

Introduction

I remember my first encounter with honey bees. Perhaps you remember yours, too. The farmer Ralph Blood, who owned the land where I grew up in southern Vermont, was a beekeeper. As part of my Boy Scout experience, I was working on various merit badges. Ralph was a Scout resource for practically everything related to animals and plants. He had a maple grove, raised cows, grew corn, kept chinchillas, and had four honey bee colonies, which he kept in the small apple orchard he maintained. He was more than happy to help a young teenager fulfill the requirements of the Boy Scout merit badge.

Ralph had a spare veil and gloves, and one day he told me to come to the farm with a long-sleeve shirt and pants and sturdy boots. He fired up a smoker using old hay from the cow barn and off we went to the orchard. He smoked the colony entrance and lifted the hive cover. Nothing happened. I looked in from the top and saw bees moving about. Ralph pried a frame free and lifted it out. I might have been a step or two farther behind, but still—nothing happened. These were not the fearsome stinging bees I was expecting. They were at home

and remained very calm as we invaded their hive. Bit by bit, I inched forward and even helped puff the smoker. Ralph cut a piece of honey-filled comb for me, and from that point on, I was hooked.

It has been said that there is more written about honey bees than virtually all other organisms, except for humans. Over a lifetime of beekeeping, which has encompassed 42 years of teaching at three different universities and 10 wonderful years of retirement, I have found this to be quite true. Despite all I know, there's always something new to learn about bees (and not just honey bees).

As a teenager, I came to understand why bees were important to us and important to Ralph. Today, bees of all kinds are every bit as important to millions of beekeepers just like him. But the future of bees is not as bright as it was back then. Certainly, Ralph must've lost a colony or two over the years. What has changed is the sheer magnitude of the loss—the large number of colonies that are dying for one reason or another. This decline is the result of a combination of stressors,

including changes to bee habitats and food sources, exposure to pesticides and more threatening pests, modern agricultural practices, and the effects of climate change. Bees are in trouble, which means that humans are in trouble, too. According to the US Department of Agriculture (USDA), bees are responsible for pollinating approximately 75 percent of the fruits, nuts, and vegetables grown in the United States. Of course, it is not only the final food product that is at risk of loss, but the beauty of the natural world, which is also directly or indirectly influenced by bee pollination. How much longer, one wonders, will we be able to stop and smell the flowers?

But there is hope. Gardeners, homeowners, individuals committed to improving the environment, hobbyists who have time and interest to devote to an interesting and fun experience—they're all doing something. They are planting flowers, flowering trees, and shrubs; they are starting honey bee colonies or erecting mason bee houses; they are committed to helping reduce the use of toxic pesticides in our environment; and they are joining together with their neighbors to save the bees.

I have assembled *The Complete Bee Handbook* in three parts:

- **Part I** will provide an overview of bees: their evolution, their present role in society, and their uncertain future.
- **Part II** will examine the most well-known bee, the honey bee—providing a cultural history of beekeeping, a glimpse into the workings of the hive, and a snapshot of modern backyard beekeeping practices; what it will not provide is a step-by-step guide to beekeeping (although I've included a few of my favorite beekeeping guides in the Resources appendix).

- **Part III** will serve as a practical resource for gardeners looking to plant flowers and trees amenable to bees and cover the products we get from honey bees—how beekeepers harvest them and the various ways in which we use them. The last chapter features 15 recipes for bee lovers, for everything from honey-centric desserts, to mead, to beeswax candles and soap.

The book is designed so that you can start at the beginning and read the book all the way through or jump in anywhere you find something of interest. Regardless of how you choose to use the book, my hope is that it will engender (or deepen) your love for bees!

PART I

The Past, Present, and Future of Bees

Where did we come from? And where are we going? I'll leave it to others to answer those age-old questions for humans. But in this part, I'll attempt to answer them as they relate to bees. In chapter 1, we'll venture back to the Cretaceous Period to learn about the origin and evolution of this incredible insect. Next, we'll fast-forward to the present in chapter 2, where you'll get a better sense of the bee's irreplaceable role in crop production. The part will conclude with a chapter assessing the bee's uncertain future—the ongoing threats to their survival and what we can all do to help save them.

CHAPTER 1

THE PAST: THE EVOLUTION OF BEES

To understand the current (and future) state of bees, it's important to first understand where they came from. Our journey begins over 100 million years ago, when the world's very first bee appeared. As we trace its path to the present, we'll discuss basic bee biology, the diverse varieties of bees in existence, and the intimate, time-tested relationship between bees and flowers.

FROM WASPS TO BEES

There is strong evidence to suggest that the world's very first bee buzzed onto the scene about 130 million years ago, in the middle of the Cretaceous Period. The Cretaceous lasted some 79 million years, starting after the Jurassic Period—the age of awe-inspiring dinosaurs, rising mountains, and receding seas—and ending about 65 million years ago.

During this time, dinosaurs and other large animals dominated both land and seas, forcing smaller mammals into the proverbial shadows—they simply couldn't compete with their larger brethren. Birds, which first appeared in the Jurassic Period, became more common—feathers helped them compete and conquer the air. The climate was relatively warm; volcanic activity was extensive. Lush forests of *gymnosperms*—flowerless plants with cones and seeds, such as tree ferns and pine and cedar conifers—blanketed the earth. These eventually gave way to flowering, seed-producing *angiosperms*, like leafy figs and magnolias, which brought vibrant colors and diversity to the landscape.

During the Cretaceous Period, large dragonflies and damselflies flew through the air. Giant butterflies prospered. Gradually, the kinds of insects we recognize today began to crowd them out. Predatory insects, like beetles, and carrion feeders, like flies, flourished with the abundance of decaying green plants and animal carcasses. So did the meat-eating wasp.

When the opportunity arose for pollen and nectar feeders to take advantage of the increasing numbers and diversity of flowers, the vegetarian bee evolved from the carnivorous wasp. In all probability, it developed from a specific wasp group known as the *Crabronids*. Crabronids establish a nest and immobilize their prey without killing them. They stock these nests with fresh spiders, grasshoppers, locusts, caterpillars, and beetles, paralyzed but still alive—no refrigeration needed.

The shift from wasp to bee did not occur suddenly. Pollen is rich in protein and would have served as a good nutritional supplement to a wasp ancestor. In an area filled with flowering plants but short on suitable victims, a few wasps began to add pollen to their reserve of paralyzed prey. Eventually, they developed more body hairs, which would allow them to search for prey in cooler weather; it would also help them collect pollen from the flowering plants. As Dave Goulson explains in his bee book, *A Sting in the Tale,* when the number of flowering plants increased, some wasps came to rely more and more on pollen, eventually cutting out animal prey from their diet entirely—and becoming bees in the process.

The Cretaceous Period ended with a literal bang. Scientists believe a giant meteor struck the area where the Yucatán Peninsula of Mexico is located today. The earth quickly became a pretty inhospitable place. Tidal waves disrupted the seas. Volcanic eruptions filled the air with so much dust it blocked the sun, causing temperatures to fall below freezing. The dinosaurs disappeared, as did much of animal and plant life. Because they could hibernate, insects and flowering plants—perhaps in the form of dormant seeds—largely survived the cataclysmic extinction event.

What Makes a Bee a Bee

So, what exactly is a bee? And precisely how do they differ from their ancestors, the wasp?

Wasps are spindly creatures, all antennae and legs. Bees, on the other hand, are fuzzy and more rounded. Wasps frequently have prominent, thin, hairless waists. In bees, the narrow body constriction of the abdomen (aptly termed a *wasp waist*) is not always obvious beneath their body hair. These hairs are the bee's prominent feature—resembling feathers under the microscope, they're ideal for trapping pollen and provide insulation during cold weather.

Pollen, along with nectar, is the staple of the bee's strictly vegetarian diet. Many wasps, on the other hand, are carnivores, consuming insects and spiders, at least as larvae. Female wasps use their stingers to inject their prey with their eggs; the prey then become an incubator for their offspring to develop. By contrast, bees only sting to protect themselves (and, in the case of social bees, their nests).

Both bees and wasps pass through a four-stage life cycle, from eggs to larvae to pupae—when the larval body transforms—to adult; social bees and wasps build a nest to protect their kin during the three developmental stages. Most bees make their nests in the ground, though about 20 percent nest in hollows in stems or burrows in trees. Development may take as long as a year or as short as two to three weeks. Adults have relatively short lives—typically a couple of weeks to a month.

It is not uncommon to find various types of bees and wasps hanging around the same flower patch, though you may not always be able to tell the difference—the most familiar wasp, the yellow jacket, is typically misidentified as a bee.

The table on page 6 highlights the distinctions between the bee and the wasp.

The Difference Between Wasps and Bees

	Bee	Wasp
BODY HAIRS	Branched, feather-like, abundant	Single, non-branching
POLLEN-COLLECTING HAIRS	Present on legs and lower abdomen	None
BODY SHAPE	Rounded	Thin and angular
DIET	Pollen and nectar only	Insects and spiders as larvae; nectar as adults
NEST	Located in the ground or hollows; constructed from glandular materials	Located in vegetation or the ground; constructed from vegetation or mud

A well-preserved bee fossil from the National History Museum in London.

The First-Known Bee Fossils

Like all animals and plants, the history of bees is written in layers of soil and rock.

Surprisingly, scientists have discovered relatively few bee fossils so far. The oldest was found in amber in Myanmar and is about 80 to 100 million years old. It displays branched hairs on hind legs, although the legs look more wasp-like than bee-like, per a 2006 report from *Science*. It's tiny, too—only about one-fifth the size of a modern-day honey bee. It resembles a group of stingless bees, an extinct species with no known close relatives.

The oldest fossil of a North American honey bee is a 14,000-year-old female worker of a now-extinct honey bee species (*Apis nearctica*). According to a paper from the *Proceedings of the California Academy of Sciences*, it was unearthed in a shale deposit in the Stewart Valley basin in Nevada, along with other insects of its day.

The oldest record of an entire apiary of honey bees was revealed by archeologists from the Hebrew University of Jerusalem in 2007. Unearthed in the Beth Shean Valley in Israel, the apiary was housed in clay cylinders and dates back 12,000 years. To this day, people in the region still keep their honey bees in clay cylinders.

A BRIEF TAXONOMY OF BEES

As early naturalists came to discover a great diversity in living things, they soon realized they required a standard classification system of some sort to keep track of them all. It became particularly apparent that there was a major division between creatures with an internal skeleton, known as *vertebrates,* and creatures without an internal skeleton, like insects, known as *arthropods* or *invertebrates.*

Though they lack a backbone, arthropods have become the most successful organisms on earth, with approximately 1 million named species (and far more unnamed). This means about two of every three known species on earth is an insect. Pretty impressive!

Within the arthropod family, insects are classified into one of 30 insect orders, the largest of which is reserved for beetles. Bees are in the order Hymenoptera, which translates to "membrane wing"—a reference to the fact that many insects in the order have clear wings. Hymenopterans are distinct from virtually all other insects in that males develop from unfertilized eggs while females develop from fertilized ones. This characteristic is termed *haplodiploidy* (try saying *that* five times fast).

The lowest common denominator of the classification system is the *species*. In essence, a species is a group of organisms that can mate and produce viable, fertile offspring. There are over 100,000 species of Hymenoptera, within which bees are grouped together as a superfamily, called *Apoidea*. There are seven families of bees and around 20,000 known species of bees worldwide. Bees inhabit every continent except Antarctica and are present everywhere insect-pollinated flowering plants occur.

The most well-known is the species of honey bee called *Apis mellifera*, originally named by Carl Linnaeus—the same 18th-century Swede who devised the entire classification system. The remaining eight species of honey bee are all from Asia, which scientists believe is the ancestral home of honey bees.

Apis mellifera has many common names, due to its numerous geographical subspecies. In the United States, we use European or Italian honey bee as the common name for honey bees, because our bees were initially imported by European colonialists, dating all the way back to 1622. A few American beekeepers keep Carniolan honey bees (a bee imported from the Carniolan Mountains of eastern Europe); in the south, the largely feral, or wild, non-managed bees (see page 28) are Africanized bees—a mixture of European and African subspecies, originally imported to South America in an effort to seek a more commercially viable bee for tropical and semitropical climates.

You Better Bee-Lieve It: The World's Largest and Smallest Species of Bee

The world's largest bee is not extinct! Formally known as *Megachile pluto*, this very special bee was rediscovered in 1981 on two islands of the North Maluku archipelago in Indonesia. *National Geographic* recently caught one on film and it quickly became a social media star. The folks at eBay say one preserved specimen sold for over $9,000.

Megachile pluto is 1½ inches long (50 mm) with a wingspan of 2½ inches (102 mm); for context, that's more than three times larger than a honey bee. It is solitary and enjoys a hearty diet of nectar and pollen, just like any other bee. One special feature is its large curved mandibles. It uses them to build a resin-lined burrow within termite nests, where it raises its young.

As for the world's smallest bee, there are two contender species on two different continents. *Perdita minima*, a ground-nesting mining bee, appears in fairly large numbers in the Sonoran Desert of the southwestern United States. According to *The Encyclopedia of Entomology*, its adult body length is recorded as 2 mm—125th the size of the largest bee. The second contender is another ground-nesting bee called *Quasihesma clypearis*, and it hails all the way from Queensland, Australia. The recorded length in *The Journal of the Australian Entomological Society* is about 1.8 mm for males and 2.1 mm for females with a wing length of just 1.2 mm. Good luck spotting one!

Solitary Bees

The vast majority of today's bees are solitary, as were most of their ancestors. Simply put, this means that other than mating with a male, the female bee works alone to rear offspring. They're quite independent, building nests and collecting nectar and pollen all on their own (although in some larger species, sisters may support one another at common nesting sites with nest digging or provisioning).

Solitary bee species tend to be small, dark, earthy colors, but a few boast a bright metallic sheen. They typically raise only a single generation of offspring per year, which the adult female accomplishes in just a couple of weeks to a month. These bees tend to fall into two broad categories, based on where they call home: ground nesters and cavity nesters.

An estimated 70 percent make their nest in the ground. Individuals dig separate tunnels in the ground, but only in a certain soil type; dry, well-drained, sandy soils are best. When soil conditions are ideal, many individual bees might use the same area while working independently. We call this *nest aggregation.*

Ground-nesting bees prepare a tunnel with salivary gland secretions to help stabilize its walls and side pockets, called cells. When they're finished, the female provisions the cell with a mixture of pollen and nectar, collected from nearby flowers. Then she places one or more eggs on the food mixture. The young will slowly mature, taking almost a year before emerging as an adult to repeat the cycle.

Cavity-nesting solitary bees, which make up about 20 percent of all bees, use naturally occurring crevices, like spaces between cracks in stones, hollow plant stems, or holes in wood, to build their nests. Mason bees, for example, line their nest cavity with neatly snipped semicircles cut from leaves, sometimes from the very plant they visit for nectar and pollen. The female bee fills these cells with food, lays one or more eggs on top, and then starts another cell closer to the entrance of the cavity, repeating the process until the entire cavity is filled. The eggs she lays in the cells closest to the entrance will produce males. They develop a little quicker and are the first to emerge, getting out of the way before the females chew their way out. Almost immediately, the males mate with the newly emerged females. The mated females then repeat the process, selecting different hollows to make their nests. How incredibly clever nature can be!

The cavity nester you're likely most familiar with is the carpenter bee. Large-bodied and colored yellow and black, these bees resemble the bumble bee (see page 68). But unlike the social, ground-nesting bumble bee, solitary carpenter bees, true to their name, tunnel into unfinished wood surfaces—often the exposed wooden beams in people's homes, barns, or wooden yard structures.

Social Bees

The remaining bees, some 9 percent, are social. Social life in insects arose through the same evolutionary processes as insect and plant diversity. Scientists believe social lifestyles evolved at least eight different times in bees, from multiple lineages. Probably the best-known social bee is the honey bee; other social insects of the same taxonomic order include ants and wasps, such as yellow jackets and hornets. Most social species sting.

The transition from a solitary life to a social one likely came as the result of either shared parental care or the construction, provision, and/or defense of a common nest; a third route to social living may have been an overlap in generations—young and adult bees living together, eating the same food.

Working together pays off: Social bees tend to live longer than solitary bees. Their nests are elaborate and well defended. Because of the size of their nest population, they tend to have an abundant, easily accessible food supply. They also have a sophisticated means of communication (more on that in chapter 5). Some are perennial, meaning they're present throughout the year, while others are seasonal, starting and finishing their social cycle within a shorter time frame.

Honey bees, in particular, are considered *eusocial*, or truly social, and are present year-round. The three components of insect eusociality as witnessed in honey bees are cooperative care of the young, division of labor with females that are highly specialized for different tasks (i.e., a caste system), and an overlap of generations. We'll discuss honey bees in greater detail in part 2.

Honey bees hovering at the colony entrance.

FAQ: Why Do Some Bees Sting?

Most of us are aware that bees sting, probably as a result of having been a victim. But let's get one thing straight: Only female bees sting. The reason is anatomical: The sting is the modified *ovipositor* of the female, meaning it helps females position their eggs. Or used to, anyway, for unlike their ancestors, bees and their female relatives (wasps, ants, and the like) no longer use their ovipositor to lay eggs. Stinging insects, like solitary wasps, use it to capture prey, while stinging social insects, like bees, yellow jackets, and ants, use it to defend their nests.

Stinging is serious business for bees, though more so for honey bees than other bees. When the worker honey bee stings, tiny barbs on the three-part sting shaft catch and hold the sting in the victim. The worker, unable to remove her sting, ends up pulling out the end of her abdomen, spelling imminent death. However, the barbs remain, pumping a powerful mixture of chemicals into the victim for at least another 10 minutes. One chemical component, an alarm pheromone, recruits other bees—it essentially "marks" the victim. Other chemicals, responsible for swelling and discomfort, are meant to impart a very important lesson: "Don't mess with me ever again!"

While all female bees (and ants and wasps) have stingers, only honey bee stings have barbs that cannot be removed once they have stung. Larger-bodied bees, like bumble bees, have larger stingers, which inflict pain but lack the chemical component that causes swelling and extended discomfort.

For information on how to avoid getting stung, see page 98.

THE COEVOLUTION OF BEES AND FLOWERS

As the diversity of flowers bloomed during the Cretaceous Period, so too did the wealth of bees. There's a simple reason for this: Pollen, the male germ cell of plant reproduction, just so happens to be the bee's favorite food. So rather than rely on the wind to spread their seed, angiosperms (plants with seeds) could now increasingly depend on bees to carry them directly from one flower to the next.

Their closely interconnected development is one of nature's preeminent examples of *coevolution*: a classic win-win situation. For every new type of flower, there evolved a new type of bee. Bees loved the pollen, yes, but they were also suckers for the flower's tasty, carbohydrate-rich bait: nectar. Producing this sugary treat costs plants, so they endeavor to hide it. Those insects and bees most likely to move pollen evolved in an "I'll scratch your back if you scratch mine" kind of way.

Today, flowering plants and insects are the two most diverse taxa on earth. The flowers come in a kaleidoscope of sizes, shapes, colors, smells, and arrangements, as do the bees who visit them. One might wonder: Why all the variation? Looking closer, it's clear that most flowers have special features designed to attract one or a small number of mutualistic pollinators.

Through coevolution, flowers and their pollinators have developed uniquely complementary characteristics. Certain flowers, for example, have adapted to be pollinated by hummingbirds. Typically, they have bright red, orange, or yellow flowers, often with fused petals, but their nectar has very little scent. Other flowers are attractive to bats; these plants feature pale, often large, bell-shaped flowers, which open only during the night. But by far, the greatest number of flowers have chosen to team up with insects. These plants have brightly colored and odorous, often abundant, prominent flowers. Among all the insects, bees are a flower's favorites.

Some flowers are generalists, benefiting from a wider array of pollinators. Likewise, some pollinators are generalists, too—the honey bee being the prime example.

Why Bees Make the Best Pollinators

Among insects, bees rule as flower pollinators. But why did bees evolve to become the predominate pollinator? Why not hummingbirds or dragonflies or mosquitos? The following section provides some answers.

All bees are strictly vegetarian. They only eat nectar and pollen, obtaining their food from a single, one-stop foraging trip to a flower. Longer tongues on some bees allow them to access nectar hidden within flowers. Shorter-tonged bees visit flowers in which the nectar is not so hidden.

Hair. Thick, branched hairs located on the bee's body are well suited to trap and hold pollen. After visiting one flower, pollen clinging on the bee's body hair may easily rub off on the next flower visited. Flowers have evolved to ensure this happens.

Pollen baskets. Bees have a special bunching of hairs to efficiently transport pollen from the flowers back to their nest. Some have such hairs on hind legs (called *pollen baskets*), others underneath the abdomen. A bee must forage a multitude of flowers to collect enough pollen to ensure that their foraging trip is energy-efficient. Before they store pollen in these special hairs, pollen is scattered over their body, increasing the likelihood that some of it will rub off on a different flower.

Bees do not eat until they get home. While bees visit flowers to collect food for developing young, facilitate normal adult glandular function, and maintain their energy, they do not eat at the flower. Rather, they move from one flower to the next before their fuel tank reaches empty, depositing bits of pollen as they go (see page 20 for more on pollination).

Bees exhibit flower constancy. When foraging for food, bees move from one flower to the next of the same kind. They don't visit a dandelion, then an apple blossom. If they require dietary diversity, they can visit another flower type in a subsequent trip; in the case of social bees, like bumble bees and honey bees, a sister in the same nest might just visit dandelions and another just the apple blossoms. By concentrating on just one flower type, the bee learns how to more efficiently obtain nectar and pollen from flowers.

Bees are specialists. Different bees specialize in different flowers, typically visiting only one or a small number of (often related) types of flowering plants. The emergence of the pollinator adult bee from its nest is synchronized for both bee and flower life cycles, boosting the odds that the bee is present when the plant flowers.

Bees enjoy sweet smells. Flowers produce and secrete nectar into the base of the flower. Nectar adds attractive smells and a sugary reward for a flower visitor—and bees just so happen to have specialized smell and taste receptors for sweet and sugary food.

Bees have perfect pitch. Flowers have special mechanisms for pollen release that only certain insects are able to trigger. Blueberries, for example, have to be vibrated at a certain sound frequency (called *sonification* or *buzz pollination*). The right sound releases pollen like a shower from the downward-orientated blossom. Bumble bees, for example, know the perfect sound pitch to release the pollen.

Size matters. Flowers blossom at the end of the plant's life cycle, and plants require fewer resources to produce small flowers than large ones. Fortunately, small bee bodies fit nicely with small flowers.

CHAPTER 2

THE PRESENT: BEES AND SOCIETY

Most people tend to think of bees as a buzzing nuisance or merely a reliable source of honey. But bees the world over are quietly going about their business, performing a significant role in our global ecology—and international economy. Because of their invaluable pollination services, bees are *keystone species*; without them, the entire ecosystem would suffer. In this chapter, we'll delve deeper into pollination and humankind's symbiotic relationship with the humble bee.

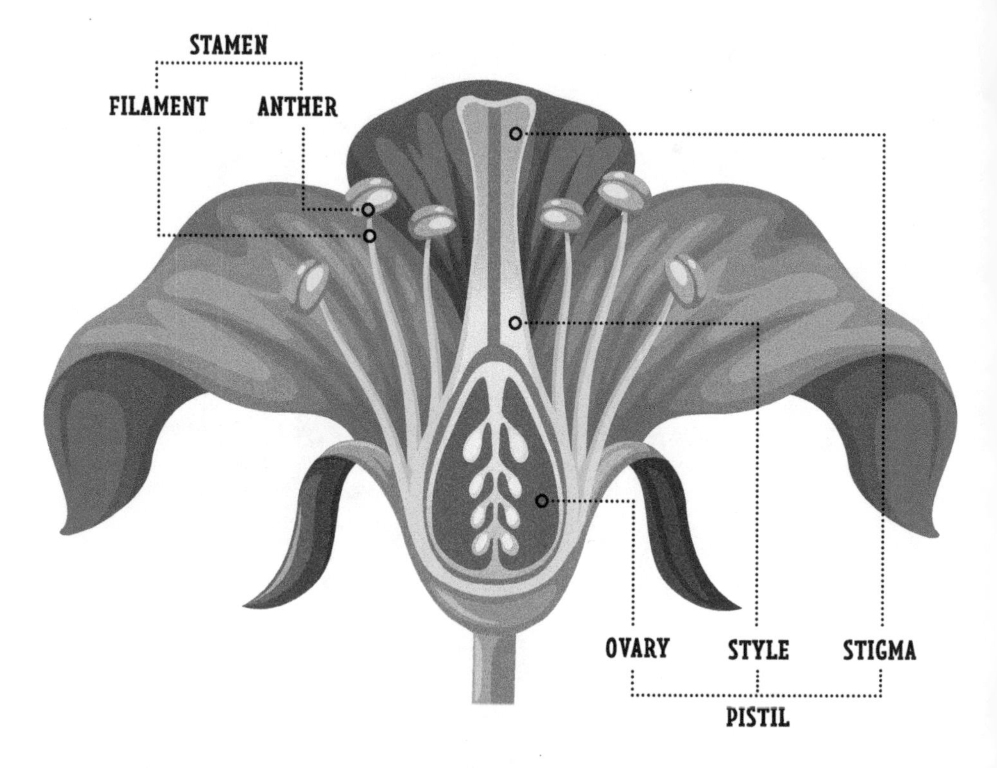

BEES ARE A FLOWER'S BEST FRIEND

As we discussed in the previous chapter, bees and flowers have perfected their lives together. Like partners in a successful marriage, the two are keen on supporting each other's success. By simply visiting flowers, bees feed themselves. But they also help feed the world. Although we might be interested in bees because they supply delicious honey (and additional products that we will explore in greater depth in chapter 8), by far the greatest value of bees—all bees, not just honey bees—is their remarkable ability to pollinate plants.

So, what *exactly* is pollination? In the simplest terms, pollination is how flowering plants reproduce. As in mammals, reproduction requires the union of female and male germ cells. The problem is that the male germ cell, a pollen grain, and the female germ cell, an egg cell, are located on different parts of the flower. The pollen grain develops on the male *stamen,* which is composed of two parts: the *filament,* a thin stalk attached to the flower; and the *anther,* the lobed tip of the filament, where pollen is produced. The female *pistil* of a flower includes a thicker central stalk known as the *style* with a sticky portion, known as the *stigma,* at its tip. To reproduce, a plant needs to transfer enough pollen grains from the anther to the stigma—in other words, it needs to be pollinated. While a number of animals, as well as a good breeze, can help unite pollen grains with stigmas, the most reliable mode of transport is pollinators, like bees.

When a compatible pollen grain reaches the stigma of the same flower species, the pollen grain moves downward within the style to reach the *ovary,* where *fertilization* occurs. Following fertilization, the plant produces seeds. Some plant seeds simply fall to the ground as the flower wilts; other flowers have mechanisms to distribute their seeds more widely. The cherry a bird might eat, for example, serves as a seed carrier. When the bird flies off to eat the cherry away from the parent plant and drops the seed into appropriate soil, a new cherry tree may grow to eventually produce blossoms, which will need to be pollinated. Wide distribution of fertilized seeds is advantageous to plants.

Fertilization and seed distribution would not be possible without pollination. Although some plants can reproduce asexually and pollen grains from the anther may fall onto the stigma of the same flower (*self-pollination*), pollen transfer by bees from one flower to another (*cross-pollination*) helps ensure diversity, bringing new traits into the next generation. Such diversity can ultimately help plants evolve to adjust to climate change, drought, or any of the hundreds of challenges they (and their offspring) may face.

As discussed in chapter 1, plants have fine-tuned flower design to facilitate pollination, with terrible-smelling nectar to attract flies; bright red, unscented, tubular flowers to attract hummingbirds; and large, bell-shaped, night-opening flowers to attract bats. But these pollinators are consolation prizes compared to the pollinator flowers most desperately want to attract: the bee.

Flowers designed to attract bees have large, prominent petals, often brightly colored, regularly and irregularly arranged around the base of the stamen and pistil; at the base, the plants have special secretory cells to produce nectar (like catnip to bees). The vibrant colors, the sweet-smelling odor, the saccharine taste of nectar, even the graceful movement of flowers in the wind—all are characteristics that attract a pollinator, especially bees. It's no wonder, then, that we see them buzzing around dandelions, lilacs, flowering apple and cherry trees, rosemary, lavender, mint, sunflowers, clovers, and hundreds of other familiar flowers.

Incidentally (or not), bees' unending appeal to plants also means they've grown less likely to self-pollinate. For example: Flowers offering pollen may be on separate plants from flowers *needing* pollen to produce seeds. Recall that during a foraging trip, bees visit many flowers but only a single *type* of flower. Their hairy bodies readily trap pollen, so when they move to a different flower, pollen grains can rub off on the stigma. Because they are not there to feed, only to bring food back to the nest, enough pollen is left for plants to reproduce.

To be fair, pollination is not something a bee does on purpose. It doesn't wake up and say to itself, "Let's go pollinate some plants." The bee visits flowers simply because that's where its food is. Cross-pollination is something of a happy accident!

Bees in the Popular Imagination: Emily Dickinson and Bees

Emily Dickinson, the famous 19th-century poet, wrote several remarkable poems singing the praises of bees. (Note: Dickinson uses *he/him*, but the bees referenced here would be female.)

My favorite Dickinson poem, "The Pedigree of Honey," is about the love affair between bee and plant.

The pedigree of honey
Does not concern the bee,
A clover, any time, to him
Is aristocracy.

"The Bee," which was apparently two poems put together posthumously, illustrates the ecstatic relationship between bees and flowers.

Like trains of cars on tracks of plush
I hear the level bee:
A jar across the flowers goes,
Their velvet masonry

Withstands until the sweet assault
Their chivalry consumes,
While he, victorious, tilts away
To vanquish other blooms.

His feet are shod with gauze,
His helmet is of gold;
His breast, a single onyx
With chrysoprase, inlaid.

His labor is a chant
His idleness a tune;
Oh, for a bee's experience
Of clovers and of noon!

The Importance of Pollination

Although environmental factors, such as wind or water, can move pollen from one flower to another, plants rely largely on an animal that flies, principally a bee, for pollination. In turn, countless organisms, including humans, rely on pollination to help produce the food they eat.

With regard to pollination, one size does not fit all. Many flowering plants have closely staked their reproduction on the needs of a small number of species—occasionally, just one. Conversely, many pollinator bees are completely dependent on specific plant species to obtain their nectar and pollen and successfully reproduce their own offspring. The ecosystemic norm is these close relationships between particular plants and pollinators.

To be sure, some plants—coincidently, many of those that produce our food—attract a wider range of pollinators or are capable of being pollinated by a generalist plant visitor, like the honey bee. Squash have a squash bee, but honey bees are rented to pollinate them; blueberries have a native blueberry bee, but bumble and honey bees are hired to pollinate them as well. The native bees (see page 30) do the best job when there are few squash or blueberry plants; the rented generalist is needed when we grow whole fields of squash or hundreds of blueberry bushes or when we grow crops beyond the natural range of their native pollinator.

The demands of modern industrial farming mean there are simply too few native bees to produce an economically viable yield. As such, the grower has to rely more and more on generalist pollinators to pick up the slack.

As shown in the feature on page 26, pollination is big business—tiny fluctuations of which can send shock waves around the world.

BEES AND THE ECONOMY

If you live in a rural area, you're likely aware that significant changes have been (and continue to be) underfoot in terms of how we grow and process our foods. Unless you're a dedicated gardener, you'd probably be hard-pressed to remember a time you ate something you yourself grew or raised. A number of factors, from innovative agricultural practices to globalization, have made it easier than ever for people to find relatively inexpensive, nutritious food. That said, it's less likely that food was produced at a small, family-owned farm near you—it may have even been produced by a farmer halfway around the world.

In addition to agriculture, beekeeping has changed as well. There has been a greater demand for generalist pollinators to supplement native pollinator populations in order to boost a

By the Numbers: The Big Business of Bee Pollination

The following statistics illuminate just how significant bee pollination is to crops in the United States and around the world.

- Bees pollinate approximately **75 percent of the fruits, nuts, and vegetables grown in the United States.** (Source: USDA, via United States Geological Survey)
- In the United States, the average daily diet consists of about **30 percent bee-pollinated plants.** Worldwide, this figure is closer to 10 percent. (Source: The National Academies Press)
- **Approximately 2.5 million honey bee colonies** are rented annually to US growers for over 90 different crops. Without pollination, 39 crops would see an immediate and drastic drop in yield. (Source: *Honey Bee Biology and Beekeeping*)
- Between **$235 billion and $577 billion** of annual global food production relies on the direct contribution of pollinators. (Source: Food and Agriculture Organization of the United Nations)
- The pollination value (not total value, just the portion of yield provided by a pollinator) is pegged in excess of **$20 billion** annually to US crop production, mainly in increased yields and superior quality crops. (Source: *PLOS One*)
- In the US, honey bees pollinate an estimated **$15 billion** of crops each year. Wild or minimally managed bees, such as bumble bees and leafcutter bees, are capable of pollinating an additional **$4 billion** of crops each year. (Sources: USDA and *PLOS One*)

farm's plant concentration and yield. For the vast majority of crops requiring or benefiting from pollination, growers contract commercial beekeepers to supply populous honey bee colonies, basically on demand. (To a lesser extent, they also ask for bumble bees, mason bees, alkali bees, and orchard bees.) With continual changes to their habitats—due to a range of environmental and economic factors—native pollinators are in peril. In urban backyards, honey bees and mason bee houses help handle flower pollination needs.

Estimates vary, but over 90 percent of the 2.7 million bee colonies under human care in the United States are kept by individuals seeking to supplement other income or by commercial beekeepers for which beekeeping is their primary source of income. Pollination rental has surpassed honey production as their main income source. In fact, 75 percent of honey consumed in the United States is imported because it's cheaper for beekeepers in other countries to produce and export to our shores.

As we have accelerated our separation from the land, colony numbers in the United States have declined. They're just half of what they were following World War II, when many farmers had bees to supply dietary sweetener (sugar was rationed at that point) and produce beeswax for munitions manufacturing. As we'll discover in chapter 6, we are in the midst of an urban beekeeping renaissance, coincident with the increasing popularity of gardening.

Wild and Managed Bees

Within the context of human interaction, we tend to break bees down into two broad categories: wild (or non-managed) bees and managed bees.

A wild bee is any one of the 20,000 species that by and large live outside the direct influence of humans. These include most native mason bees, carpenter bees, bumble bees, mining bees, and many more—90 percent are solitary as opposed to social.

But here's where it gets a bit more complicated. A small number of quote-unquote "wild bees" are, in fact, managed—albeit in a very limited sense. In the case of the introduced leafcutter bee, a mason bee, we prepare shelters and stock them with cartons of straws or blocks of wood with holes drilled in them. For the blue orchard bee, another mason bee, we provide hollow reeds. Beekeepers can disassemble these nests, clean them of parasites and diseases, and use them to restock an area in a subsequent season.

Beekeepers also partially manage the wild bumble bee. They capture or raise queens from hibernation, provide boxes and nest materials, and feed them so they'll grow colonies. Then they move growing colonies into greenhouses to pollinate tomato plants or set them in fields to pollinate blueberries, watermelons, and the like.

Collectively, it is hard to put a value on the pollination service of wild bees—and scientists are constantly trying to find new ways to bolster both minimally managed and truly wild bees. Some evidence, summarized in a 2019 paper in *Frontiers in Ecology and Evolution*, suggests that wild and managed bees may not always be compatible. Suffice it to say, our world would be much more drab, and our food selections less diverse, without their services.

That said, we would be equally lost without the honey bee, the pollinating workhorse that is the most intensely managed bee. Technically, the honey bee is not native to the Americas, though it's been here for so long, few people would identify it as a foreigner. Some would, however, consider it domesticated livestock under the general definition of "domestication": a mutual relationship with humans, who provide care and facilitate reproduction. Bees are still able to escape their human caregivers and survive in the wild. These escaped, wild, non-managed bee colonies are termed *feral*. By some estimates, there are as many feral bee colonies in the United States as there are managed.

The means by which honey bees escape human management is via swarming (see page 91). If beekeepers fail to capture the swarms that leave their managed colonies, the swarm may move into a tree hollow or a cavity in a man-made structure or under the overhang of a cliff. These feral colonies also swarm, ultimately establishing their own feral nests. Just because a colony is feral, though, doesn't mean its members stop pollinating flowers—making them every bit as valuable to the environment as their managed kin.

WE NEED THE BEES

Farmers whose crops require cross-pollination—the fruit orchardist, the herb farmer, a sunflower grower, those cultivating pumpkins or cucumbers or watermelons—all of them need bees. And if they're not keeping bees themselves, they must contract with commercial beekeepers to get them.

When farms were small and diversified, bee colonies were often part of the farm. The growth of industrial agriculture (see page 37) has meant farms around the world are bigger than they've ever been. They rotate a smaller number of crops—or even just plant a single crop. As such, they need to be treated more intensely with pesticides; if one pest or disease slips by, it could threaten the entire crop.

The Almond Board of California's annual reports offer an illustrative example of just how much farmers are reliant on bees—and how supplying those bees places much strain on beekeepers and bees.

You Better Bee-Lieve It: Bees Make Your Favorite Foods

About one-third of the foods we regularly eat depend upon bee pollination. It provides the color and variety in our diet. Here are some popular native North and Central American foods pollinated by bees, compiled with research and statistics from *Crop Pollination by Bees.*

BLUEBERRY/CRANBERRY: Although there are native bee pollinators, such as the blueberry bee, commercial production is courtesy of rented honey and bumble bees.

SQUASH/PUMPKIN GOURDS: Native bees, such as the squash bee, are specialist pollinators. But as field size increases, growers must supplement with honey bee rentals to satisfy demand for Halloween pumpkins. All those pumpkin-spiced drinks and foods use the artificial essence of pumpkin.

AVOCADO: This popular fruit has very complex pollination needs. Individual trees produce an abundance of flowers. Honey bees are required for commercial production, but stingless bees pollinate avocados in Mexico and Central America.

LIMA (AND PINTO, NAVY, BLACK) BEANS: Yields of these legumes are improved by up to 30 percent when bees are present for pollination. They also produce a watery-white, highly fragrant honey. Lots of native bees visit beans.

TOMATO: Bumble bees are necessary in greenhouse production. Otherwise, employees often "pollinate" flowers with a vibrating toothbrush.

CHILE/BELL PEPPERS: Although these are self-pollinating, bee visitation increases fruit size and provides earlier and more uniform ripening and a more symmetrical pepper. Sonification, primarily by bumble bees, helps improve cross-pollination, given the wider distribution of pollen released when they shake flowers.

Almonds are a medium-size tree that grows well in the Mediterranean climate of California's Central Valley. Nearly half of the world's almonds are grown there. Because almonds are wholly dependent on pollination, yields are practically nonexistent without bees. According to bee economist Brittany K. Goodrich, there are 1 million acres of bearing trees currently, each acre of which requires one to two honey bee colonies for an economically sustainable yield. For the early season bloom of February and March, these farmers require the rental of 1.8 million of the estimated 2.7 million total colonies maintained by US beekeepers. This number, however, is far more than the beekeepers of California can annually maintain on a year-round basis. So, for this bloom period, two-thirds of the colonies are transported to the almond orchards from the Midwest, Texas, the Gulf Coast, and even as far away as New England.

For beekeepers, the fee to supply bees to almond farmers is outstanding—double or more what apple growers or blueberry farms are willing to pay later in the season—and way far more than they get for honey. But there is a major downside for the beekeeper: Colonies, even those coming from the deep southern states of Texas and Florida, are only beginning their spring expansion in February. To meet the minimum colony size, beekeepers need to artificially boost colony strength by feeding considerable amounts of sugar and protein to their bees, starting in early January. The movement to almond farms contributes to colony stress. Due to the higher concentration of colonies in one place, bees from one colony may very well be exposed to the pests or viruses of bees from another.

Most of us are isolated from what is happening in agriculture. But worldwide there is an expansion of pollination-dependent crop planting. Ask most kids where milk comes from and the answer is "the milk carton"; ask them where apples come from

and the response is "the grocery store." Those same youngsters may know that honey comes from bees. What they're less likely to know is that the alfalfa the dairy cows graze on was pollinated by bees or that the apple orchardist rented bees to ensure they had enough harvestable fruit to supply the supermarket.

CHAPTER 3

THE FUTURE: SAVE THE BEES

In the previous chapter, we discussed the importance of bees, both to the ecosystems in which they thrive and to the US and global economy. In this chapter, we'll explore the uncertain future of bees and identify several major threats. I promise it's not all doom and gloom! We'll also discuss what is being done to protect the bees—and what you can do as an individual to improve the state of bees.

NATURAL-BORN BEE KILLERS

All living organisms have to deal with a number of pests and pathogens—some major, others minor. Bees are no different. Yet with their elaborate nests, stores of food, and many offspring, they're an appealing target for enemies.

There are numerous perpetual threats to bees. Soil-nesting wild bees must contend with fungi and harmful bacterial diseases as well as nest mites, ground beetles, and pestiferous ants. Parasitic bees and wasps can also trick them into rearing their young on the food stored in their nests as opposed to their own offspring.

Social bees, like bumble and honey bees, have more elaborate nests, masses of provisions, and plenty of developing young brood (in egg, larval and pupal life cycle stages), all of which are coveted by a number of other organisms. Even the beeswax combs they fashion within the nest can end up ransacked.

Honey bees, in particular, historically face four big pathogenic diseases: *American foulbrood* and *European foulbrood,* both caused by tiny bacteria; *chalkbrood,* a fungus that infests and kills developing brood; and a microsporidian fungus called *nosema,* which resides in the adult bee gut and robs them of nutrition.

There are a host of minor diseases as well. Beekeepers seeking to keep healthy bees need to learn how to recognize disease symptoms and when and what to do to take corrective measures to protect their bees if necessary.

Additionally, a number of pests may lurk around the beehive, seeking entry or waiting for victims to emerge. Fortunately for honey bees, they've got guard bees to keep enemies out (see page 84); naturally, the bad guys still find ways to target vulnerabilities and gain access. When bees leave the shelter of their nest, a host of pests are ready to make them their next meal; ants, wasps, predacious bugs, and flies catch bees as they exit the colony or while they are outside the home foraging. Crab spiders and other predators camouflage themselves atop flowers and patiently wait for bees to visit.

Although social bees have always been susceptible to viruses, they've become more threatened as of late. Some viral infections may reach epidemic proportions and cause a colony to collapse (see sidebar on page 42). We lack a bee vaccine to combat virus epidemics, so beekeepers need to focus on preventive measures, like reducing stress of pathogens and making sure bees have an adequate diet. The search for locally adapted bees, ones that are better able to resist viruses as well as harmful pests and pathogens, is ongoing.

In general, monitoring for presence and prevention is the best strategy to ensure bees remain healthy. Early detection gives beekeepers more options to keep pest and/or pathogen populations at bay. If runaway harmful pest or pathogen populations are taking over, beekeepers may use targeted control measures; in these instances, nonchemical options—like changing the position of the bees' home, setting a trap for a particular pest, or even genetically changing the bee population to become more resistant—are preferable to pesticide or antimicrobial chemical methods.

NEW AND ONGOING THREATS

On top of the natural-born bee killers discussed in the previous section, there are numerous socioeconomic and environmental factors that continue to pose a threat to bee populations.

Together with *pests* and *pathogens*, *pesticides* and *poor nutrition* make up what apiarists typically call the 4 *P*s. We can add a fifth *P*, too—*people*, beekeepers among them, who lack an understanding of the value of bees and who, inadvertently (or worse, deliberately), harm bees. Because our food production (and ecosystem in general) depends on bee pollination, we critically need solutions—not just for the range of established pests and pathogens, but for new and ongoing threats.

I've outlined the big ones in the pages that follow, as well as what can be done to mitigate them.

The Varroa Mite

The Problem: The vampiric mite *Varroa destructor* literally sucks the life out of bees—and is currently the most serious threat to honey bees in the United States, Canada, and Europe, according to a 2019 report in the *Proceedings of the National Academy of Sciences*. These tiny, eight-legged bee pests live harmoniously on the Asian honey bee, *Apis cerana*. But when beekeepers from Europe brought their bee colonies to the Philippines, the mite unexpectedly changed hosts. The mite is a menace to European bees, which lack the ability to keep mite numbers under control the way their Asian cousins do.

Unfortunately, the story of the varroa mite is an all-too-familiar one these days. Despite border inspections, the nature of global trade is such that pests and pathogens

A phoretic varroa mite on the thorax of a worker bee. These mites transfer to developing workers just before the pupal stage.

frequently, if accidentally, gain access to a new host, which unsuspectingly introduces them to an entirely new region of victims. Colonizer pests and pathogens initially have a biological advantage over existing pests and pathogens. Varroa mites immediately became a significant concern for honey bees in their new home.

The Solution: The great rise in backyard beekeeping could be a possible solution, as it could aid in finding and selecting bees better suited to combat mites. Backyarders keep bees differently and have the time and inclination to approach bee care differently than do those who need to make a living from bees. Scientists and hobbyists have suggested that the pool of backyard bees could supply the genetic breakthrough they need to find a better, safer solution to varroa mites—namely, breeding bees that are resistant to virus epidemics.

Industrial Agriculture

The Problem: When we talk about industrial agriculture, we're typically referring to the increased mechanization that goes into growing, harvesting, and processing our food. Just as we love our "smartphones," farmers love their "toys." If you've visited the countryside lately, you may have even noticed how large their tractors have become. Such equipment is expensive, so it's necessary to farm a lot of land to pay for it. Rural county roads used to be lined with small family farms; nowadays, there might only be one or two.

Farming is widely considered a wise investment. Consequently, foreign investment is on the rise, and fields are increasingly owned by equity firms, large corporations, or conglomerates. To buy or rent land these days, you have to have more money and bigger machines to stay above water. Federal and state government policy looms large over farming practices;

farm profit or loss can depend on whether you receive a government subsidy.

Again, it's worth asking: What effect does this have on pollinators? Ultimately, beekeepers need locations where their bees might have less exposure to pesticides and an abundance of flowers to forage. Fewer landowners actually operating their farms mean less direct, hands-on land stewardship, particularly if the new owners have less connection to and experience with farming.

The Solution: The need for supplemental pollination is growing, not diminishing. Growers, beekeepers, pesticide companies, and other supply chain manufactures need to sit down and develop rational strategies to protect native bee populations and ensure the health of honey bees. They need to take these concerns to federal and state governments and press for implementable legislation.

Federally, there is a set-aside land program wherein the government pays farmers *not* to farm fencerow to fencerow—to guarantee wetlands and natural areas are "set aside" and not simply filled in and plowed over. Set-aside strips of flowering plants—not crops, but flowers grown exclusively for pollinators—have been shown to improve pollination without detracting the pollinators' value to the crop bloom. It's in a farm's best interest to host bees. Such options exist today, but certainly need to be expanded.

Monoculture

The Problem: In the simplest terms, monoculture means growing a single plant at a time. This happens a lot in the United States. Just look at the corn belt of the Midwest, where field after field is filled with corn, or the Palouse region of eastern Washington, where wheat covers the rolling hills as far as the eye can see. Monoculture is the agricultural practice of choice in industrial agriculture—particularly for large companies whose technological advances have made it possible to grow crops in places they could not have in the past.

Monocultures are an economy of bigness and of sameness. They allow farmers to increase efficiency and yields in a competitive marketplace. But they also expose their crops to greater risk from pests and pathogens than would crops grown in a polyculture; as a result, they necessitate a stronger reliance on pesticides.

Bees fare better in diverse environments for numerous reasons. Chiefly, a monoculture provides pollen and nectar only if and while it's flowering. Preceding and following the bloom period, however, monocultures are deserts, offering nothing for flower visitors; whereas a polyculture may offer bees pollen year-round, a monoculture means its feast (during bloom) or famine (during the rest of the growing season).

The Solution: Why should the vast bulk of the government subsidy programs go to the large industrial farms with their monocultures? The big agricultural lobbyists who push for these subsidies obviously have a different opinion, but it's clear that expanded programs for organic or sustainability growers, who still farm with crop and animal diversity, should be fairly subsidized as well. These growers demonstrate every day that pest and pathogen control can be integrated into best management practices; natural controls may be a bit more complicated and labor-intensive, but they can work. We can grow cheap food, raise healthy animals, and save pollinators—all at the same time.

Pesticides

The Problem: When farmers concentrate on one type of plant in an area, any pest or pathogen that can utilize that plant for its food suddenly has a gigantic plate to eat off of. As such, using a pesticide to kill weeds, soil-dwelling insects, birds that descend upon a field to feast on seeds, or yellow jackets that puncture and drink the sweet juices of fruit has become a common practice when farming vast acreages of corn and other monocultures. Nowadays, they're so cheap and accessible, we use some of them simply as a preventive measure—just in case the pest might show up. In fact, it's nearly impossible to buy seeds of some common crops that are *not* treated with a pesticide. Centuries ago, Native Americans planted corn to have four ears—one for the birds, one for the insects, one for the plant fungi, and the last for themselves to eat. That probably worked well back then. But imagine what our foods would cost if we allowed pests and pathogens to take three-quarters of the harvest each planting season.

Ultimately, pesticides should be used as a last line of defense, not the first. Agricultural chemicals are tested for negative effects on pollinators; if they can potentially harm pollinators, their registration label must state as much. However, the tests may not account for every potential danger. Long-term effects, or accumulated exposure over time, may not be readily apparent.

The favored pesticide du jour is always changing. Pests and pathogens are also quite adept at developing resistances to chemicals, so older materials eventually become less effective (or a cheaper alternative enters the market). Company representatives aggressively push their preferences. And their chemists are continually seeking "safer," more effective pesticides.

One such pesticide is *neonicotinoids*. First introduced in 1985, they have, since the early 2000s, become some of the most widely used pesticides for killing insects. Yet they've also become the most controversial.

Environmentalists consider neonicotinoids highly poisonous to pollinators and stream and soil insects, as well as a major contaminant to soil and water. Studies gathered in a 2016 report from the Xerces Society show that *neonics* (as they're informally called) shorten the life of honey bees, make honey and bumble bees forgetful, and are a likely cause of reductions in feral bee populations, especially ground nesters and bumble bees. The European Union has banned their use on crops that attract pollinators, and several US states (like Vermont and Maryland) have passed restrictions on sales to untrained applicators; other states are considering doing so. Yet thus far, the US Environmental Protection Agency (EPA) considers the benefits to outweigh the risks. (Xerces.org has continually updated information on how neonicotinoids kill bees.)

The Solution: Outright bans on pesticides are not going to work. They will just become contraband. There are risks with pesticide use to be sure, but there are also benefits. Achieving the right balance is not always easy. Restricting the most dangerous pesticides, to pollinators and ourselves, to those trained in safe pesticide use is common sense. Training programs are mandatory but need to be beefed up with respect to pollinator protections. Currently, point-of-sale companies are limiting sale of neonicotinoids to the general homeowner because governments have generally failed to do so.

Manufacturers of pesticides should be provided incentives to develop newer, safer pesticides that are thoroughly vetted for potential harm to pollinators before releasing them to the public. It costs millions of dollars to bring new chemicals to markets. The costs of ensuring their safety should be paramount in their development.

By the Numbers: The Decline of Bees

Recent studies report widespread declines in insect populations around the world, including bees and pollinators. What's not clear is the extent of the declines—and where, exactly, they take place.

Surveys of native bee populations generally lack reliable conclusions that speak to the extent of wild bee decline. The losses are more inferred than specifically documented. But the evidence that has been found is compelling. Of 187 New York bee species represented by preserved specimens in Cornell and American Natural History Museum insect collections, 49 are currently in decline and considered rare and/or potentially threatened, according to a 2013 article published in the *Proceedings of the National Academy of Sciences*.

The decline of bumble bees, our best-known native bee, is also cause for concern. According to a recent analysis by the Xerces Society and the IUCN Bumble Bee Specialist Group, a whopping 28 percent of North American bumble bees are considered at risk of extinction. Three species are considered gone entirely; one, the rusty patched bumble bee, was officially listed as an endangered species in 2017.

The story of decline is a bit more complicated for honey bees: Annual surveys of US beekeepers, conducted by the Bee Informed Partnership for the past 13 years, suggest an average percentage of colony loss across all 50 states as 38 percent, up significantly from the 12 prior year colony loss average of 28 percent. In my independent survey of Pacific Northwest beekeepers for the past 10 years, I found that the 40 percent loss rate of backyard beekeepers doubles that of the commercial beekeepers. In total, honey bee colony numbers are now about half of what they were estimated to be following World War II.

There is a bright side: Other estimates point to increasing numbers of bee colonies worldwide. The global honey trade is cited as largely responsible for these increases, as is the increased cultivation of pollination-dependent crops. A 2009 analysis of Food and Agriculture Organization (FAO) data published in *Current Biology* suggests managed honey bee hives have increased as much as 45 percent worldwide during the last half century. Even declines in Europe and North America have been reversed. The USDA's latest estimate of colonies in the United States, which had dipped to 2.4 million in the wake of the destructive varroa mite, has now increased to 2.7 million.

A bumble bee approaches a flower.

Climate Change

The Problem: Although some politicians may disagree, the science is clear: Humans have had a significant impact on the warming of our planet. Climate change doesn't merely mean higher temperatures, higher rain and snowfall, or more pronounced natural disasters, like hurricanes and tornados. It also means higher sea levels, longer and hotter summers, changes in the normal moisture availability of soils, and a host of other environmental disruptions related to food production.

So, what does all this mean for the bees? Because flowers and bees evolved together, it's only natural that what affects one will likely affect the other. Soil-nesting bees need specific soil types; additional moisture or hotter temperatures might adversely affect their nesting areas. According to a 2020 study published in *Science*, bumble bee nesting has been shown to be negatively affected by volatile temperature fluctuations. And if flowers do not last as long due to changes in the climate, they likely won't be able to offer pollen or secrete nectar for as long, either.

One illustrative example comes courtesy of the great naturalist Henry David Thoreau, who recorded the flowering dates of 500 plants around his cabin in Walden Pond and home in Concord, Massachusetts, during the mid-1800s. When Elizabeth Ellwood and other researchers updated the phenology records (phenology is the study of timing of plant events, such as flowering) with flowering dates from 2004 to 2012, they found that Thoreau's plants, on average, were flowering 11 days earlier than during Thoreau's time; in 2010 and 2012, the region's two warmest springs, they found that flowering was occurring a good three weeks earlier.

Walden Pond is not an isolated incident: When the researchers updated phenology records from the notes the great American conservationist Aldo Leopold made at his isolated "cabin" along the Wisconsin River, similar discrepancies were found. These findings are consistent with those in areas across the country.

The Solution: Climate change is going to necessitate tough solutions. It starts with those of us on the local level being willing to make changes. But we also need accurate information and informed politicians who are willing to make the tough decisions. Politicians seldom look beyond the next election. Adapting to climate change requires long-range vision.

For bees, in particular, the EPA and Food and Drug Administration (FDA) need to be de-politicized and empower the dedicated employees—working on better and safer foods, pollinator protection, cleaner water, reduced air pollution, and ways to mitigate the inevitability of climate change—to make a real difference.

THE MYSTERY OF COLONY COLLAPSE DISORDER

In the early aughts, beekeepers the world over witnessed the disappearance of their colonies' adult bee population. The bees vanished, without their typical cluster overdeveloping brood. The abandoned colonies still had stored food reserves. But with too few workers to support them, the queens were unable to function. In two to four weeks, the colonies collapsed.

A new term, *Colony Collapse Disorder* (CCD), was coined in 2006 to describe this phenomenon. To this day, the cause or causes of these sudden and widespread colony losses remain a mystery. Beekeepers typically expect a loss of 10 to 15 percent of colonies annually—particularly during the overwintering period. But according to *Geography Compass*, the losses have become two to three times greater.

There are any number of possible culprits for colony collapse. Records show a history of sudden localized loss events, given names like "autumn collapse," "spring dwindling," and "fall colony decline." But most were usually localized in impact, and often did not show up again in subsequent seasons.

What has changed that might be so negatively affecting bee health? Are honey bees losses a canary in the coal mine—a potential wake-up call for our own species that the ecosystem we share with honey bees is untenable? Can we learn something by better understanding what is killing the honey bee?

Investigators have spent years seeking answers for colony collapse. Factors identified as likely responsible for the high levels of colony losses, ranked in order of significance, include the invasion of the varroa mite, bee exposure to chemicals meant to kill pestiferous insects but that inadvertently also kill beneficial honey bees, and a shrinking food base less capable of sustaining adequate bee colony health. (We can consider poor colony health a combination of these three major factors.)

Beekeepers who require healthy bees to rent to neighboring farmers for pollination and consumers looking for clean, chemical-free honey and other bee products are pioneering novel controls in their efforts to keep their bees productive. They monitor and evaluate their colonies constantly, integrating control approaches to keep the mite population below levels that would cause economic harm.

To be fair, bees were never an easy animal for beekeepers to husband. But whatever's causing colonies to suddenly collapse is not making those efforts any easier.

You Better Bee-Lieve It: Insect Drones and Artificial Pollination

Drones—not to be confused with the drone bees in a honey beehive (see page 79)—have a bright future in agriculture. They can fly over crop monocultures and quickly detect weed or pest infestation hot spots so farmers can deploy spot treatments without having to spray the entire field. Got a leak in irrigation system? No problem for drones. They save the cost of laborers having to walk the field to search for the problem area. Nutritional deficit in a crop? A drone can quickly and accurately assess problems. Bottom line: Drones help ensure healthier crops with higher yields for the farmer.

Other kinds of drones are starting to prove their worth as pollinators. These high-flying wonders are especially tiny. They can be equipped to carry freshly harvested pollen and function like a bee by rubbing against exposed anthers and transferring some of the pollen.

Government funding is being used to further develop these drones for artificial pollination. The Department of Defense Advanced Research Projects Agency recently supplied a multimillion-dollar grant to develop "RoboBees." According to a report published in *American Entomologist*, "The current version weighs less than two grams, beats its wings 120 times per second and can perch, fly, and swim." Unfortunately, "it still can't navigate around other flying objects"—which means it may run into some issues with the living creature it's designed to replace.

HOW YOU (YES, *YOU*) CAN HELP SAVE THE BEES

When it comes to saving the bees, there's no lack of information. And that's for good reason: We need the bees! Information abounds on websites, blogs, and flyers and in magazines and books like this one, offering advice about what we might do to help save the bee. The three major suggestions typically involve improving bee habitat, reducing pesticide use, and providing ample nesting areas for bees.

I encourage you to look at the checklist for gardeners included in chapter 7 (page 120). In addition, here are five practical things you can do for bees:

1. Start a beehive and learn how to manage it to ensure overwintering success. If that is not feasible, encourage cavity-nesting bees, such as mason bees, to visit your garden by buying or building bee houses.

2. Support your local beekeeper by purchasing local honey and/or beeswax. The local (artisanal) honey is better for you, anyway.

3. Even if you're not a serious gardener, plant more flowers, reduce lawn size, and use pesticides sparingly.

4. Become a Bee Ambassador. Communities designated "Bee Cities" and schools that have become "Bee Campuses" (a Xerces Society program) need individuals to take their conservation message to the community. (For more, see the Resources appendix.)

5. Get down with the cause. Assist local, regional, and/or national groups that support bee habitat improvement. Three of my favorite nonprofits include Project Apis m., Honey Bee Health Coalition, and Xerces—the single unique invertebrate conservation society.

PART II

All About Honey Bees

In part I, we took an expansive view of bees, tracing their evolution into the present and beyond. In part II, we'll zoom in on the most beloved (and important) species: the honey bee. We'll start in the Stone Age, where fearless honey hunters braved bee stings to secure sweet honey. As we make our way to present day, we'll take a peek inside the hive to uncover the fascinating machinations of a honey bee colony. Finally, I'll provide you with a primer on backyard beekeeping.

CHAPTER 4

A SHORT AND SWEET CULTURAL HISTORY OF BEEKEEPING

The relationship between humans and bees predates recorded history. The motivating factor, undoubtedly, was quite simple: *honey*. Long before sugar cane came along, honey was the most accessible sweetener. In this chapter, we'll examine the cultural history of beekeeping—sometimes described as the second-oldest profession in the world—from mankind's first interactions to the present.

HONEY HUNTERS OF THE STONE AGE

Do you remember your first encounter with a honey bee? Maybe you can recall tasting honey or have a memory of being stung? Perhaps you had a beehive visit courtesy of a friend or family member? For many of us, our earliest bee-related "memories" are the stories our family members tell. (I talk about mine in the Introduction.)

Just as we, as individuals, may be uncertain about when we first came across a honey bee, we, as humans, have no *real* way of knowing when we first encountered bees. Most likely, it was to collect honey; almost certainly, our ancestors paid the price and got stung. How did they know that a bee nest contained sweet, delicious honey? Perhaps it was from watching a chimpanzee push a stick into a tree hollow with bees and withdraw it to lick the dripping honey. Or maybe they saw how honeyguide birds chattered and preened to lead ratels (honey badgers) to bees in Sub-Saharan Africa, waiting to collect their reward in honey scraps after the badgers destroyed the nests. Humans, too, found they could follow the honeyguides—some even do so to this day.

The oldest record we have of a bee nest harvest is a detailed rock painting dating back 8,000 years found in the Cuevas de la Araña—or Spider Caves—located near Valencia, in eastern Spain. It depicts an individual with a basket, suspended on vines on a cliff face, reaching into a bee nest with bees flying about. The person does not appear to have any protection to avoid being stung.

The earliest humans ate what they caught. They lived in small bands or extended families, led a migratory existence, and had to hunt for food almost daily. It is unlikely that honey hunting was a major daily activity for them, though it's hard to imagine they passed up the opportunity when it arose. When these honey hunter-gatherers did seek out honey, they targeted feral bee nests for their high-protein brood and energy-rich honey despite the threat of stings and the inherent risk involved with accessing nests high in trees, within cavities, or beneath overhangs, perched high on barely accessible cliffs.

Needless to say, hunting wild bee nests was not an easy job. But the rewards outweighed the risks. Seeing as sugar would not become widely available until Chinese and Indian traders learned how to cultivate it around 1200 to 1000 BCE, honey proved a prized dietary sweetener—and a valuable bartering item. Beeswax was every bit as valuable, as it was used for everything from waterproofing clothing to sealing stored foods and as a source of light.

Today, honey hunting has mostly faded into obsolescence. In Africa, the practice of humans following honeyguide birds is nearly lost, given the dearth of badgers and a more settled human population. Generally speaking, keeping bees in handy containers, known as *rustic hives,* close to a family residence has obviated the need for cultures in rural areas worldwide to locate nests in the wild.

That said, some cultures still gather honey the old-fashioned way. In the Himalayas, a *National Geographic* photographer recently braved stings from the giant honey bee *Apis dorsata* to film Gurung honey hunters of Nepal scaling sheer cliffs to harvest its single-comb feral nests. A smallish number of Gurung still return annually to an area where feral bee nests are known to occur to rob them of beeswax comb, bee brood, stored pollen, and, of course, honey. As an article for Oregon State University details, a particularly daring group of honey hunters still operates in the Sundarbans, a vast mangrove swamp spanning a swath of the India–Bangladesh border, contending with saltwater crocodiles, Bengal tigers, and giant boa constrictors in order to steal from feral *Apis dorsata* nests. We should consider ourselves lucky for being able to purchase honey from a farmers market!

A recent book has sparked something of a revival among hobbyists looking to employ this once-valuable skill. In *Following the Wild Bees: The Craft and Science of Bee Hunting,* Cornell University bee expert Professor Tom Seeley has reintroduced an age-old bee nest hunting technique, which essentially involves caging several foraging bees, feeding them honey, and then following them as they make a beeline to their nest. He has used it for pioneering research to understand how feral bee colonies remain healthy and productive in areas where managed bee colonies experience heavy annual losses.

THE FIRST BEEKEEPERS

As time progressed, many cultures transitioned from hunting and gathering to growing food and caring for animals. In his epic treatise of human development over the past 13,000 years, *Guns, Germs and Steel,* evolutionary biologist and noted ornithologist Jared Diamond observes that the transition was not immediate—nor did it occur simultaneously in China, the fertile crescent of the Middle East, and Mesoamerica, the three independent birthplaces of much of the plants and animals we cultivate today.

Raising crops and husbanding animals did not necessarily make for an easier life. The earliest farmers continued to hunt and harvest from nature (as rural peoples still do) while gradually increasing dependence on cultivated grains, fruits, and corralled animal meat and products (such as milk).

We can visualize a similar evolution relative to honey bees. As the human population grew, so too did advents in farming and technology. Hunter-gatherer populations were pushed to the edges of this rapidly developing world. The need to bring bee nests closer to human residences was necessary for bee products, such as honey, to remain important to humans.

Moving bees closer initially meant keeping bees in their original nest cavity. Eventually, they used materials at hand to

Bees in the Popular Imagination: The Enduring Mythos of the Honey Bee

The great ancient Western cultures have a rich mythology regarding bees—one recurring theme in all of them relates to the origin of bees.

- **In ancient Greece, Aristaeus was the god of beekeeping, honey, and mead. Aristaeus is blamed for causing the death of Eurydice, wife of legendary musician Orpheus, because she stepped on a snake while attempting to escape him (only in Greek mythology . . .). Legend has it that the Eurydice sisters took revenge on Aristaeus by killing his bees. Aristaeus let the bee bodies rot in animal carcasses, from which live bees emerged to fill his emptied hives. (Source: Arista Bee Research Foundation)**
- **In ancient Egypt, burying an ox was the reported method of getting bees. It was believed that live bees would emerge from the base of the horns.**
- **Samson, in the Old Testament Book of Judges, describes a different animal origin for bees: "Out of the eater, something to eat; out of the strong, something sweet" (Judges 14:14, *NIV*)—a reference to his finding honeycomb in the carcass of a lion.**
- **In ancient Rome, the poet Virgil devotes most of the fourth book of *Georgics* (translation: *Agriculture*) to honey bees. He speaks of the life and habits of bees as models for current Roman society. He also reports that bees are spontaneously born from the carcasses of oxen.**

Surprisingly, all of these myths may have had *some* basis in fact: The rib cages of decaying large animal skeletons could provide a suitable cavity for a bee nest in arid, tree-less habitats.

build nests; wood or clay or straw containers enabled far easier and greater harvests. According to *Beeswax Alchemy*, the first beekeepers kept clay hives over 8,000 years ago in Egypt. Honey, while still a unique sweetener, became a valued natural medicine—an increasingly covetable commodity in more densely crowded areas. Along with beeswax, honey remained a valuable bartering item, sought by other community members in exchange for surplus plant crops and animal services.

In the sections that follow, we'll take a look at typical bee farmers of three great Western civilizations.

Ancient Egypt

The Egyptians are frequently credited as the first beekeepers. Bees were closely connected with Egyptian royalty, even if they were (perhaps unsurprisingly) kept by poorer farmers. For at least 8,000 years, the Egyptians have used clay tubes and pipes to house bees. These hives have an entrance for bees on one end and are harvested from the other. They are typically stacked 10 or more high, forming giant triangles.

The Egyptians were also the first *migratory* beekeepers. They transported their cylindrical clay hives on special rafts, seasonally moving up or down their great river of life, the Nile. At appropriate sites, they took the hives off the rafts and into the fields. There are indications that Egyptian farmers may have developed a rudimentary understanding of the significance of pollination.

Honey played an important role in Egyptian culture. Traditionally reserved for royalty, all social classes used honey eventually. It was a sweetener, a medicine, and even a currency; people used it to pay taxes. In her book *The Sacred Bee in Ancient Times and Folklore*, Hilda Ransome notes that one marriage contract from the time read, "I take thee to wife . . . and promise to deliver to thee yearly twelve jars of honey."

Honey was also an embalming fluid, and tombs of Egyptian royalty often included jars of honey reserved for the afterlife.

When the famous archeologist Howard Carter discovered King Tut's tomb in 1922, he found that it contained 2,000 jars of honey. Because honey doesn't spoil, it was still perfectly okay to eat. (Apparently, the king had not yet used it in his life everlasting.)

Ever resourceful, the Egyptians made good use of beeswax as well—to make candles, to seal foods in containers, and to mold figurines, some of which were weaponized as amulets to cast spells on their enemies.

Ancient Mayans

History has puzzled over Spanish records regarding the Mesoamerican Maya and Aztec populations' use of honey—as a sweetener but primarily as a medicine. There were no honey bees in the Americas at the time of European conquest, during the 16th century. It turns out the Mayan and Aztec peoples obtained honey from *Melipona* bees, the social stingless bees, known locally as *señoritas*.

Represented by over 500 species, the *Melipona* genus is very diverse. Stingless bees live only within the equatorial regions around the earth. Species vary in size, from smaller than fruit flies, to those larger than the honey bee. According to archeologist K. Kris Hirst, the Mayans favored the honey bee-size *Melipona beecheii,* whose colony size varies from a few hundred to thousands of workers.

We do not know when the Mayans moved from merely harvesting stingless bee honey from forest trees to keeping bees in their farms and gardens. It was undoubtedly a gradual process. Their growing population and increasing distance to forest nests may have been a factor. Bees and honey were considered sacred. Perhaps as society stratified and priests became nest guardians and harvesters, they found the need to keep bees closer to the temples. Perhaps bees were used to pollinate vanilla orchids or Capsicum peppers—two staples of the Mayan diet.

Or perhaps it was always about the honey. To this day, stingless bee honey is used to treat eye, ear, nose, throat, digestive, and skin issues. However, as a piece published on the archeology platform Ancient Origins observes, these traditions are unfortunately being lost among Mexico's current-day Mayans, given the miniscule harvests and the increased dominance of Africanized bees.

Ancient Romans

Rome was founded in 753 BCE and lasted well over 1,000 years. Throughout their reign, Romans conquered and ruled much of Europe, northern Africa, and the Middle East—regions where agriculture of grains, olives, and grapes predominated. They adopted and improved upon many practices from the people they conquered—including beekeeping. Roman beehives included Egyptian clay pipe hives and straw skep hives of the lowlands of the North Sea region; there were also tree cavity hives in regions where forests predominated.

Most farms in the Roman Empire (27 BCE to 476 CE) began as small and subsistent. But they were increasingly purchased by the church or wealthy individuals, who transformed them into larger land tracts. The same farmers still worked the farms, but as sharecroppers or slaves. These farmers tended a panoply of crops and husbanded a diverse array of animals. Because many of the crops did not provide nectar or pollen, bee colony holdings were minimal. When beehives were harvested, they were largely destroyed—the bees were driven off, and the combs were removed and crushed to separate honey from beeswax. Honey was bartered for other services or transported to the larger cities for the nobility.

Great Roman authors wrote extensively about bee culture, including Cato the Elder, in his treatise *De agri cultura* (translation: *On Farming*); Columella, in his 12-volume *De re rustica;* and Pliny the Elder, in his encyclopedia *Naturalis Historia.*

Honey Around the World

The ancient Egyptians, Mayans, and Romans weren't the only ones to endow bees and honey with cultural significance. People all over the world developed customs around this special insect.

ABYSSINIA (MODERN-DAY ETHIOPIA)	In the old Ethiopian empire, the groom had to bring a quantity of honey to his prospective bride before the wedding. If the amount was unsatisfactory, the bride and her family rejected him as a future husband. (Source: *The Beekeeper's Bible*)
EUROPE (WALES, IRELAND)	According to 19th-century lore, hives need to be told of important events in their keeper's lives, especially the death the of owner. If the bees are not "put into mourning," they might leave, not produce honey, or die. (Source: *Telling the Bees*)
BABYLONIA (IRAQ) AND GREECE	The term *honeymoon* comes from ancient moon-time cycle cultures. Newly married couples drank mead (honey wine) for its supposed aphrodisiac properties during the first moon cycle (first month, the sweetest) of their marriage to ensure lots of offspring.

FERTILE CRESCENT (PRESENT-DAY IRAQ, IRAN, SYRIA)	"Tanging" is banging on a tin pot while chasing a swarm (see page 91) so the bees will land nearby and be captured. According to Virgil, it was thought to permit the owner to chase a swarm invading a neighbor's property while announcing ownership of the bees in flight.
ENGLAND	An old English custom documented in *The Honey Makers* dictated that if you sell bees, you must receive gold as payment. But it is considered unlucky to begin keeping bees by buying them—you must await a swarm to select you as their owner.
LITHUANIA	According to Lithuanian folklore, if during the first evening spent in the husband's house a bee stung the new bride, the bride was considered to be a poor wife and a lousy choice.
BULGARIA	On February 23, the feast day of the patron saint of beekeeping, Saint Kharlampii—a renowned healer who is said to have used honey and beeswax in his strictly natural remedies—families would bake hive-shaped pies. Orthodox church priests would bless honey and ask it to cleanse the land of pestilence and plague. (Source: Druidry.org)
EGYPT	Ancient Egyptians believed the tears of the Egyptian sun god "Re" transformed into bees when they reached the earth, providing Re's followers with liquid gold, or honey. (Source: *The Tears of Re*)

FROM BEE FORESTS TO BOLES

Throughout the Middle Ages, feral honey bees lived in tree cavities in much of Europe's forests (as many still do today). Tree hollows are generally small, dark, sheltered cavities with a single entrance, making them convenient for bees to defend. This entrance could be closed to bee size with bee glue—otherwise known as *propolis*, the resin collected from tree sap and other plants—especially with the approach of cold weather. Inside, the bees coated interior walls with propolis (for its antimicrobial properties) and suspended their parallel beeswax combs from the top of the cavity. The combs did not extend all the way to the bottom of the cavity, where dead bees, wax, remains of brood, and colony debris accumulated, enabling scavenging flies, beetles, and small microorganisms to break down and recycle nest waste.

The characteristics of tree hollows that made them ideal for bees were duplicated when humans started to make their own hives. Humans initially hunted feral bee nests in the forest, probably in conjunction with hunting animals, and looking for ramps, mushrooms, or other forest foods. They marked trees to claim the bees as their property. At the appropriate season, they would return to the site and cut into the tree to remove honey-filled combs. Although the harvest destroyed the combs, the bees would not be killed. Closing the cavity and leaving some brood-filled combs behind allowed the surviving bees to replenish their combs and owners to return the next season to re-harvest honey.

Harvesting often meant a long trip into the forest. At some point, the bee hunter realized that the hollow section of the tree containing the bees might be moved to a more convenient location on his farm. Or better yet, he could simply hollow out a log and entice bees with a tantalizing offer of beeswax comb after rubbing the interior with sweet-smelling flowers. *Baited hives*, as these were called, are still used today to capture swarms searching for a new home.

Where trees were scarce or were considered the property of royalty (recall how the Sheriff of Nottingham pursued Robin Hood's band in Sherwood Forest), bees were kept in *skeps*. The icon of a beehive, skeps are conical-shaped, tiered hollows fashioned from straw or reeds. Stone walls included bee *boles*, indentations with a shelf, to hold skeps. The straw might be "clomed"—covered in mud and dung—so it weathered better. Skeps were the first hives colonialists introduced to the New World.

Bee skeps

The Most Important Moments in Modern Beekeeping

THE EUROPEAN HONEY BEE TAKES OVER THE WORLD

Over time, beekeepers began housing hives in boxes in which the combs hung from the cover. Boxes were more sensible to carry and made it easier to cut the combs to harvest. But that all changed in the early 1850s, when Philadelphia minister L. L. Langstroth, housing his bees in wooden champagne crates he'd picked up from the city docks, accidentally raised his hive cover and discovered that his bees had not glued the top of the comb tightly to the top of the box. Immediately grasping the significance of this, he set about devising a hive that respected *bee space*—three-eighths of an inch, the height of a bee—around the combs, between boxes, and beneath covers within the hive interior so he could enter and freely manipulate the combs. He patented his hive in 1852. It was the first hive to feature practical, removable combs that beekeepers didn't have to cut out to harvest. Although European and other American beekeepers were not far behind in developing their own removable comb hives, Langstroth's design prevailed. To

1609

Charles Butler, author of *Feminine Monarchie*, one of what will become many beekeeping books, is the first to correctly identify the large hive "ruler" as the queen bee.

~1735

Carl Linnaeus (Carl von Linné) names the honey bee of Europe *Apis mellifera* (honey maker) and then tries, unsuccessfully, to change the name to the more accurate *Apis mellifica* (honey gatherer).

CONTINUED >

this day, we still call our beehives "Langstroth hives" (see photo on page 64).

By the 19th century, lumber for boxes and frames was more readily available, roads had improved, and the ability to transport colonies was advancing rapidly. Beekeepers, in addition to seeking the best domicile to house the bees, sought to improve the tools they used to keep bees. Two important ones, the smoker and hive tool, were improved (see page 97 for more details). An extractor to produce liquid honey without crushing comb was developed. The foundation of beeswax with embossed worker cells was perfected. These innovations and many others all came about before the turn of the 20th century.

While the technology was changing, so too was the beekeeper's choice of honey bee. In North America and Europe, the most popular honey bee was northern European—the black or German bee. But this bee can be temperamental and has difficulty with European foulbrood (see page 34). Reports

1852

L. L. Langstroth, a Philadelphia minister, discovers the "secret" of bee space; produces first movable frame hive.

1860*

L. L. Langstroth successfully imports first Italian queens into United States.

from the valleys of Northern Italy started reaching American beekeepers, espousing the virtues of a different bee—the Italian bee. It provided large honey surpluses and was apparently not susceptible to European foulbrood. The larger beekeepers of New York, Vermont, Pennsylvania, and Ohio all sought to import Italian queens. Langstroth finally did so successfully in 1860.

Over the next 50 years, other bees were imported as well, such as the Carniolan bee (across the Adriatic Sea from Italy), the Caucasian bee (from eastern Europe), the bee of Egypt, a bee from Syria, and bees from the Mediterranean islands. Naturally, other hives were patented, too, and the right way to manage bees was fiercely debated at national and regional meetings. There were two competing viewpoints: One held that honey should only be harvested in the comb (pure honey, just as stored by bees); the other held that beekeepers should use a larger hive, grow bigger colonies, and extract the liquid honey

1861

The first issue of *American Bee Journal* is published. It's the oldest English language beekeeping publication in the world and has been published continuously since 1888, except for a brief period during the Civil War.

1865

Franz Hruska develops the first practical centrifugal honey extractor.

CONTINUED >

for greater harvests. This second view would come to dominate the expanding *apiculture,* or bee culture, industry.

But in the end, it was Langstroth, his hive, his book about how to manage bees, *The Hive and the Honey Bee,* and his advocacy for Italian bees that had an everlasting effect on modern beekeeping.

WHAT ABOUT THE HUMBLE BUMBLE BEE?

While they're two of the most recognizable social bees, it would be difficult to confuse a bumble bee (in England, often called the "humble bee") for a honey bee. Bumble bees are among the largest bees. Their bodies are generously covered with bright, bold, distinctly yellow and black hairs. They are the cold weather bees of the northern hemisphere (hence their hairiness). And unlike honey bees, they do not produce honey for us to harvest.

~1884*

Moses Quinby of New York becomes the first commercial beekeeper. He completes a transfer of bees from box hives to movable frame hives.

1889

George M. Doolittle of Onondaga, New York, develops a comprehensive system for rearing queen bees. It's still in use today—and is named the Doolittle Grafting Method.

Bumble bees differ, too, in that they're only seasonally social. They are not around during the winter. Fertilized females (queens) emerge in the spring from winter hibernation. Initially, queens must do all the work. The first priority is to find a suitable nest site in the soil. But they don't dig, so they look for an abandoned vole nest in a field, an unoccupied mouse nest, or someplace where the soil has collapsed, forming a cavity. In our gardens, they might settle at the decay spot in the middle of the compost bin, on the underside of some stones, or beneath a landscaping timber or bale of hay.

The emerged queen improves the nest space and builds a series of waxen cells for raising some babies and tiny pots to store a bit of ripening nectar. She lays a few fertilized eggs and then busily collects nectar and pollen, food to feed her developing larvae.

1907*

US beekeeper Nephi Miller transfers his hives to areas around the country to increase their productivity during winter. Since then, migratory beekeeping has become widespread in the United States.

1927

Karl von Frisch, an Austrian behavioral scientist, introduces his dance language communication (see page 87) to initially skeptical scientists. He was awarded the Nobel Prize in 1973.

CONTINUED >

Numbers depend upon the species, but initially, she rears only a half dozen offspring. When they reach the pupal stage, they are enclosed together (not individually as in honey bees) in a waxen envelope. These need 28 to 35 days to become the first workers, depending on how warm she keeps them.

With the first workers, the queen completes her working phase and becomes the egg layer. Given favorable weather, the nest continues to grow, although most species do not ever have very populous nests. At the end of the season, males and larger females are produced. The large females will be queens in the following year. Males and the entire worker population die with the approach of cold weather, while the mated females seek a hibernation shelter to pass the winter.

2006

Dennis van Englesdorp coins the term Colony Collapse Disorder (CCD, see page 45) to explain syndromes of sudden losses of bee colonies. Excessive colony losses continue in North America and Europe, although beekeepers find ways to replace lost colonies.

2014

President Barack Obama creates the Pollinator Health Task Force to investigate the issue of disappearing bees and other pollinators.

***Dates taken from** *Beekeeping in the United States.*

CHAPTER 5

THE HIVE MIND

In the previous chapter, we discussed the cultural history of beekeeping. Now, let's delve into the hive itself. We'll trace a year in the life of the hive, identifying who does what and exploring the honey bee's incredible means of communication. Whether you're interested in beekeeping or actively beekeeping already, this chapter will expand your knowledge of and fascination with the honey bee.

WELCOME TO THE HIVE

As you'll recall from the previous chapter, early honey hunters looked in trees or caves for nests to rob. As humans began to more closely husband honey bees, they used containers that were convenient, such as log hollows, hives constructed out of rolled bark or mud baked into clay, large clay pots, woven reeds, or simply hives made out of wood. Until Langstroth incorporated bee space, bees suspended beeswax combs from the tops of these cavities.

Following Langstroth's example, modern beehives are constructed of wood, although plastic is increasingly becoming the material of choice. At the most basic level, the hive must serve as an enclosed cavity. Sizes of the wood or plastic boxes can vary but are more or less standardized to approximately 20 by 16 by 10 inches. Inside the boxes, wooden (or plastic) frames hold beeswax combs, where the nectar ripens into honey and pollen into *bee bread* (a protein-rich food for the bees). In other combs, they raise their young.

Traditionally, the beehive entrance is at the bottom, usually built into the bottom enclosure. The opening size can be narrowed with a wood, plastic, or metal reducer. For convenience and to help the materials last longer, the hive sits on a stand of some sort. The covers are separate and often include two pieces. The inner hive cover may be fabric; plastic is generally avoided because it will trap moisture inside. During spring and fall, it may be replaced with a sugar water feeder.

The single piece of furniture inside the beehive box is the frame. You might think of this frame the way you would the frame of a painting, except instead of holding your treasured piece of art, it holds the bees' treasured wax combs. Bees mold these combs into hexagonal cells, back-to-back, with a three-part cell bottom shared with cells built in the opposite direction. The frame gives beekeepers a way to hold and examine the combs without damaging the combs themselves, which are delicate and easily crushed.

To help ensure bees build their comb within the frame outline—something they don't necessarily always do—a template sheet (or foundation), embossed with the base of the hexagonal cells, is securely positioned in the frame center. Adult bees produce wax from abdominal glands and build cells outward from the template. This action is called *drawing comb*; when they're finished, it's called a *drawn comb*.

Frames are suspended on a ledge cut into opposite sides of the box, and when completely drawn, the outer edge of the cells (the open end) varies only by bee space from the adjacent comb. Not only do frames protect the combs, they are expressly designed to maintain bee space around all sides of the drawn combs, including space between additional boxes added on top of the original hive box.

When bee space is respected, the bees will not build additional comb between adjacent frames, between the outside frame and the hive wall, or between boxes (though they will draw some comb as ladders to access additional boxes). If the space is too large, the bees make extra comb to fill it; if the space is too small, bees will not be able to move freely, and they will close the space with propolis.

Critically, bee space allows beekeepers to freely remove the frames and inspect the hive. We owe Langstroth big-time for this discovery, so it's no wonder his legacy lives on.

A Year in the Life

A hive is designed to house the total colony population. The size of the colony will vary from season to season. Beekeepers will add boxes with additional combs as the colonies grow and even more boxes (termed *supers*) to house surplus honey; as winter approaches, boxes will be removed.

From a low winter population, pollen from spring flowers incentivizes queens to lay eggs, and the population of worker bees increases until the colony reaches its maximum size. These numbers will contract come fall, as the weather begins to turn and resource availability declines.

It may be helpful to visualize the annual cycle as a bell-shaped curve, with the lowest population during the winter period, growth in the spring, peak numbers in the summer, and consolidation in the fall. The population of brood—the number of beeswax cells filled with developing young—follows the same seasonal pattern.

During the annual cycle, the colony will have one queen. More often than not, colonies will raise some additional queens and then swarm in late spring, prior to peak population (see page 91 for more on swarming). Male drones will be reared in the spring but then curtailed after the population reaches its peak. Any surviving adult drones are kicked out of the colony in the fall. The curve is compressed into six months in the far north and elongated to cover all 12 months in the deep south. Generally, after peaking at 50,000 to 60,000 workers, colonies will drop to 12,000 or fewer workers in their dormant period.

The length of buildup and peak population size is directly related to nectar and pollen resources available to the bees and weather conducive to foraging (see page 85). With more resources, the queen is stimulated to lay more eggs, provided sufficient comb space is available, leading to more brood-rearing—which, in turn, stimulates greater growth.

FAQ: How Exactly Do Bees Stay Alive During the Winter?

The dormant phase of the honey bee's annual cycle, winter, can be cold and long up north or less cold with only a few parka-worthy days in the deep south. Bees do well at these extremes and everywhere in between. Where winters are cold and long, the bees hunker down, clustering closer and closer together. But how exactly do they survive? The answer gets to the very heart of why humans keep bees: honey.

When there are flowering resources, bees collect nectar and ripen it into honey. They do so to guarantee themselves an energy supply that will last throughout winter. We refer to winter as dormancy, but that's actually a bit of a misnomer. Bees do not hibernate the way a bear crawls into her den and enters into a deep sleep, surviving on fat reserves in her body. While the fall bees do increase body fat reserves, they don't have the luxury of a bear's long sleep. Remarkable as it may sound, they remain active all winter. Their secret is to cluster together for warmth, eating honey stores that supply just enough heat to survive.

During this dormancy they will begin to rear brood, who will grow up to replace the colony's aging bee population. Compared to the summer bee, who literally works herself to death in six weeks, the winter bee suspends its aging process, living six months with relatively little activity.

It should be said that bees do not heat the whole cavity of the nest or hive—merely the cluster portion around their queen. As winter progresses, the cluster slowly moves upward to remain in contact with honey-filled cells. The cluster forms at about 50°F. As temperatures drop, chilly bees on the outer surface move inward toward the center. On the coldest nights, the cluster might grow to be the size of a soccer ball.

MEET THE HONEY BEE FAMILY

Social insects, such as honey and bumble bees (see page 68), and wasps, such as yellow jackets and hornets, operate under a caste system, meaning that female adults are specialized to perform different tasks. In honey bees, a single queen is the reproductive female. The workers, her daughters, are specialized to work—to do the various tasks necessary to maintain the home, store the food, protect the colony, and more.

With that in mind, let's lift our proverbial frame and take a closer look at what's going on inside a typical hive.

The Queen

As a separate caste, the queen is easily recognizable and distinguished from the workers. She even looks different, with an elongated abdomen (housing the egg-producing ovaries), simplified mouth parts, fewer body hairs, and no barbed stinger or hindleg pollen basket. She will live one to three years—far longer than her daughters. As she is the only egg-laying female, bees pay her special attention, and beekeepers take pains to avoid injuring her.

The single bee colony queen has two functions: lay eggs and produce her unique pheromones. These chemicals, produced in internal glands, serve as the social glue that unites a colony and enable workers to recognize their own individual queen. Each bee family has its own distinctive odor, resulting from the queen's distinctiveness and the smells of the foods they collect. Collectively, this is called *colony odor.*

Queens move slowly and methodically about the beeswax comb, seeking empty cells in which to deposit a single egg. Sometimes she is fussy and won't lay her eggs because the cell has not been thoroughly cleaned and polished by workers from a previous inhabitant. When inspecting the cell, she perches on the open end and, using her front legs like a pair of calipers, measures the opening size. The majority of cells will have an opening about one-fifth of an inch in diameter; inside, she will lay a fertilized egg. If cells are larger, about one-fourth of an inch in diameter, the egg passes from her body without being fertilized. While unfertilized eggs in much of the animal kingdom simply do not develop, in bees and other hymenopterans, they develop into the male.

The Drones

Drones, the males of the bee family, are the brothers of the workers and sons of the queen. They, too, are specialized. They don't do queenly things, nor do they work. Their specific task is to mate with newly emerged adult queens—to pass their sperm onto the queen. Unfortunately for the drone, successful mating also means immediate death. The process of withdrawing his mating parts after discharging his sperm into the queen ends up ripping open his abdomen, and he falls paralyzed to the ground. Some success, huh?

Drones make up a relatively small percentage of a colony population. At peak season they usually number a few hundred up to 1,000; by comparison, there are several thousand workers. Drones are larger than workers and stouter, barrel-shaped. The abdomen is more solid and less striped. A distinguishing feature is their huge eyes—the drone head is mostly the two compound eyes. And, of course, as males, drones lack a stinger.

Drones never mate inside the colony. To accomplish their one task, they leave home in afternoon hours and fly to specific congregating areas—called *drone congregation areas* (DCAs)—within a mile or so of their home. At a DCA, they form large, flying comets, composed of 1,000 or so drones. Those who don't find a mate may live a month, except in the fall, when drones are thrown out of the hive entrance as the colony completes its fall preparations to overwinter.

The Workers

Workers, the second female caste, constitute the vast majority of beehive inhabitants. Unlike the queen, they are unable to mate. As explained on page 75, the numbers of workers fluctuate depending on the season.

Workers need approximately three weeks to progress from fertilized egg through the larval feeding stage to a capped stage, where individual pupa are encased in the cell with beeswax covering the outer, open end. The queen and her workers pass through the same three developmental stages, but queens develop faster (16 total days) while workers take an additional five days before emerging as adults. Drones take the longest (24 days) to develop—perhaps that's why drones are considered so lazy.

Once workers reach the adult stage, they pass through a sequence of duties, starting out as a house bee before transitioning to a field bee and foraging outside the hive, collecting nectar and pollen. For worker bees born in the spring and summer, the adult will live for about six weeks; for worker bees born in the fall, the adult may live up to six months, depending on the severity and length of the overwintering period.

What follows are several brief descriptions of the common hive and field tasks workers perform, from newly emerged adult to old age.

HOUSE BEES

Newly emerged adult workers take some time to get accustomed to their surroundings. They engorge on pollen, helping activate glands needed to perform later tasks. One of the house bee's earliest duties is cleaning cells, starting in the brood nest, the warmest region of the hive where it was born. House cleaning involves removing debris, smoothing over the remains of the wax cappings on the outer cell rim, and polishing interior cell walls.

If the queen happens to come by, house workers will join her retinue. They feed the queen, remove her wastes, and groom her body, all the while picking up her pheromones. As retinue workers meet and greet their sisters, they distribute these pheromone chemicals, which communicate that all is well and to keep working because the queen is present.

NURSE BEES

After keeping house, the young worker becomes a nurse bee. Nurses feed the developing bee larvae. Each developing cell is checked on an average of 1,300 times a day, approximately every three minutes. Hungry larvae are supplied with *worker jelly,* a nutritious blend of highly digestible food manufactured by hypopharyngeal glands in the nurse bee's head.

As the developing larva ages, its diet changes to a mixture of bee bread, nectar, and honey, taken from nearby storage cells. The number of days nurses spend tending the brood depends on the amount of brood in the hive and the urgency of other competing tasks, but a typical shift lasts a week or two.

NEST PROVISIONING BEES

After a few busy days feeding larvae, worker bees switch to nest provisioning. This entails taking nectar from returning foraging bees and retreating to a quiet place in the hive to "work" it, thoroughly mixing in the enzyme *sucrase*. Then they place droplets of the nectar along cell walls to create large-surfaced bubbles. The enzyme inverts (or breaks down) complex nectar sugars, and the hive atmosphere absorbs moisture from the droplets. As ripening and inversion continue, droplets are consolidated.

The nest provisioning bee also processes the pollen the foraging bee directly deposits into comb cells into bee bread by mixing glandular secretions and nectar into it and pressing it into cells. When cells are three-quarters full, they put a honey glaze over the top. Lactic acid fermentation and chemical changes convert the raw pollen into bee bread within the storage cells.

UNDERTAKER BEES

In addition to house cleaning, worker bees have to remove the bodies of bees that die in the hive as well as the remains of brood and any intruders. If an intruder is too large—like, say, a mouse—the worker will encase its remains within propolis so its putrid odor doesn't "contaminate" the hive. Propolis has antimicrobiological properties and is additionally used to coat the interior of the hive to help protect hive occupants from microscopic pathogenic enemies too small to be seen.

WAX WORKERS

At around 12 days, worker bees have functioning wax glands. The wax comes out as a thin flake. Using its legs, the bee passes several flakes to its mouthparts, where it mixes saliva to mold the flakes together. The beeswax is used to cover (or cap) the worker pupae or fully ripened cells of honey to keep contaminants out.

Sometimes bees need lots of beeswax to build parallel combs, such as when a swarm moves into a hollow cavity. The bees chain, or festoon, together, clinging to sisters by outstretched legs, facilitating wax gland activity and mouthpart molding of the wax flakes into beeswax. In two or three days they can construct two to three parallel combs, extending downward three to five inches, thus kick-starting the process of filling their new cavity.

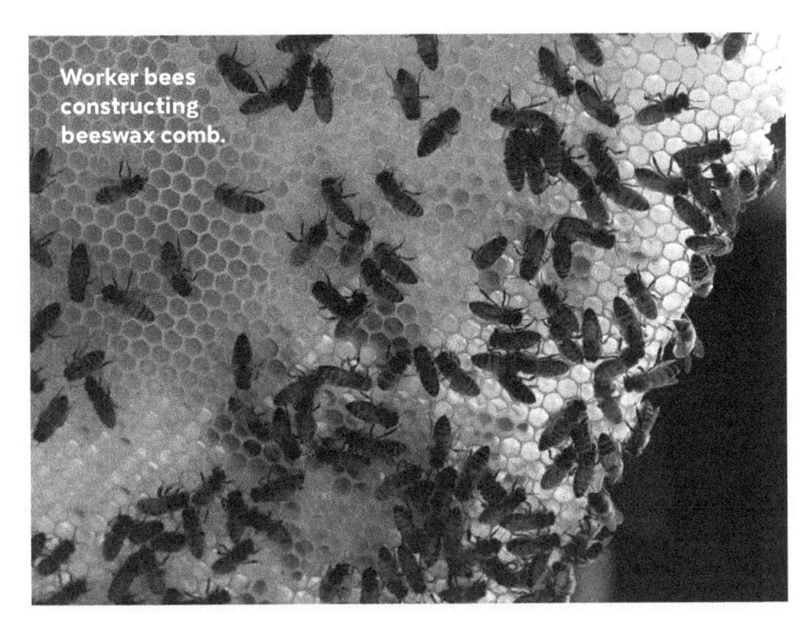
Worker bees constructing beeswax comb.

GUARDS

The honey beehive is a veritable treasure chest of highly valued resources. To protect their bounty, honey bee workers have the modified ovipositor, the stinger, to protect themselves and their resources. Bees with fully developed sting glands become hive guards.

Guarding is usually the last of the worker's household chores. Guards are positioned at the hive entrance. They stand gallantly in a state of readiness, elevated on middle and hind legs, with front legs and antennae actively moving before them. They spot different behaviors and check for body odor. Only family members are allowed to pass, as they have a familiar smell. Potential enemies are challenged and physically blocked from entry or grasped and dragged away from the entrance. If intruders persist, the worker might sting an intruder. When the sting is employed, an alarm odor gland releases a special pheromone chemical that serves to mark the enemy and recruit other bees in the vicinity to help expel the intruder.

FORAGERS

As worker bees age, they begin to venture outside the hive to begin orientation flights. First, they just fly immediately in front of their home. Soon after, they fly tentatively around their home, eventually widening the flight arc to learn landmarks and the exact location of their hive. By about three weeks of age, the worker is ready to become a *field bee*—in other words, a bee that travels outside its home. She will likely collect (or *forage*) nectar and pollen the remainder of her life, surviving perhaps another three weeks. She might also collect water or propolis.

The colony is totally dependent upon foragers to collect the resources they need. But the worker bee's final duty is likely its toughest. Foraging is dangerous work. Field bees can run out of food, get lost, get picked off by a bird or predacious insect, or simply get blown off course and be unable to properly return home. Moving in and among plants, the wings of field bees take a beating. (In fact, you can recognize older foragers by the amount of wear and tear on their wings.)

SCOUTS

A subset of field bee, fundamental to the hive's success, is the scout bee. These are the worker bees tasked with helping a swarm (see page 91) find a new home site and helping the colony maintain a diverse diet.

Scout bees are more numerous in the morning hours and less numerous when the bees find a major source of nectar to exploit. When pollen is scarce or there is a lot of brood in the colony, more worker scouts will seek additional pollen sources. Scouts may be bees who leave home just to explore, or they may be foragers on a waning or not-so-rich source looking for a new option. These scouts may forage on a source that closes at midday (such as watermelon) and, instead of staying home in the afternoon, search for alternative flowers.

Like human families, bee families also have members who are not content to follow the beaten path. The scouts represent the future of the hive, as they're always looking for what else might be out there.

WAGGLE DANCE

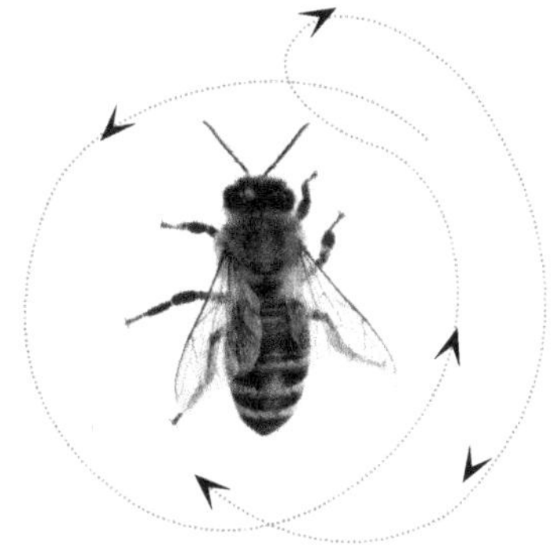

ROUND DANCE

THE DANCE OF THE HONEY BEE

Humans communicate with one another primarily through words. Honey bees, on the other hand, dance. Like a couple performing the waltz, forager and scout bees repeat a series of no-less-graceful movements. But they don't do so for show. Rather, their dances are a means of sending quite detailed messages to their hive mates.

Two such dances exist, both of which are performed for one of two reasons: to recruit foragers to food or to flag a new potential home site for a swarm (see page 91).

For nearby flowers (or home sites), the bees perform a *round dance*. The bee moves in an exaggerated, sped-up circular pattern and then repeats the same circle in the opposite direction. This dance sends the message that food (or a new home site) is close by. The Italian bee uses the round dance for sites within 40 yards, while German bees—studied by Martin Lindauer and his supervisor, Dr. Karl von Frisch, who pioneered the research on dance language—uses the round dance to indicate food or home sites within 88 yards (80 meters). Like humans, different bees have different dialects.

When food (or a new home site) is farther away, however, more precise information is needed. Enter the *waggle dance*. Distance is communicated with a 250-hertz frequency sound emitted during the center waggling portion of the bee's dance, at which point the dancer vigorously moves her abdomen from side to side. (You may do something similar with your own abdomen on the dance floor.) The follower bee times the waggling sound segment, which is then converted to the distance a bee would have to fly once she left the hive.

Because bees cannot simply point to where to fly from inside the dark hive, the dancer also gives precise directions. She does this by orienting her body position relative to gravity during waggling. After a straight waggling run, the bee turns right, returning to the same gravity orientation, and then waggles again (with sound production) before turning left and circling back to resume the same body orientation. She will repeat this over and over. Each complete dance cycle resembles a figure eight.

When the recruit has followed the dancer through several cycles and timed the waggle portion, plus gotten its taste of the nectar and smell of the flower pollen clinging to the dancer's body, she retrieves some honey for fuel and heads over to the colony entrance. There, she recognizes the polarized light pattern of sunlight. Transposing the gravity angle she picked up following the dancer, she orients her outgoing flight into a sun angle; an orientation away from gravity tells her to fly directly toward the sun's position, while an orientation toward gravity tells her to fly away from sun's position. With remarkable accuracy, she will find the dancing bee's site. Way to go, bee!

Dancing is one of the bee's most fascinating capabilities, permitting foragers to concentrate their energies on the richest, most profitable flower sources and scouts to inform a clustered swarm about a new potential home site cavity.

The casual observer can witness their dances when a comb is removed from a bee colony and put behind glass in an observation hive. Amazingly, humans with nothing more than a stopwatch and a protractor (to determine the gravity angle) can watch them dance and learn to speak "bee."

You Better Bee-Lieve It: Royal Jelly, Your Highness?

Some call it the miracle food. And, indeed, for a developing bee, *royal jelly* can mean the difference between becoming a sterile worker and a new queen.

Royal jelly is a glandular food that nurse bees produce in the pharyngeal glands of the head and thorax. At first, it's fed to all bee larvae. Later on, it's given exclusively to developing queens.

To raise a new queen, workers construct special, vertically oriented queen cells. Larvae in these cells are fed with copious amounts of royal jelly, which triggers internal changes within the body of queen pupae that do not occur in workers or drones. Consequently, the queen adult will emerge with the fully developed ovaries and pheromone glands she needs to rule the hive.

In the photo of uncapped cells below, the royal jelly is the shimmering pool of liquid in which the larvae rest.

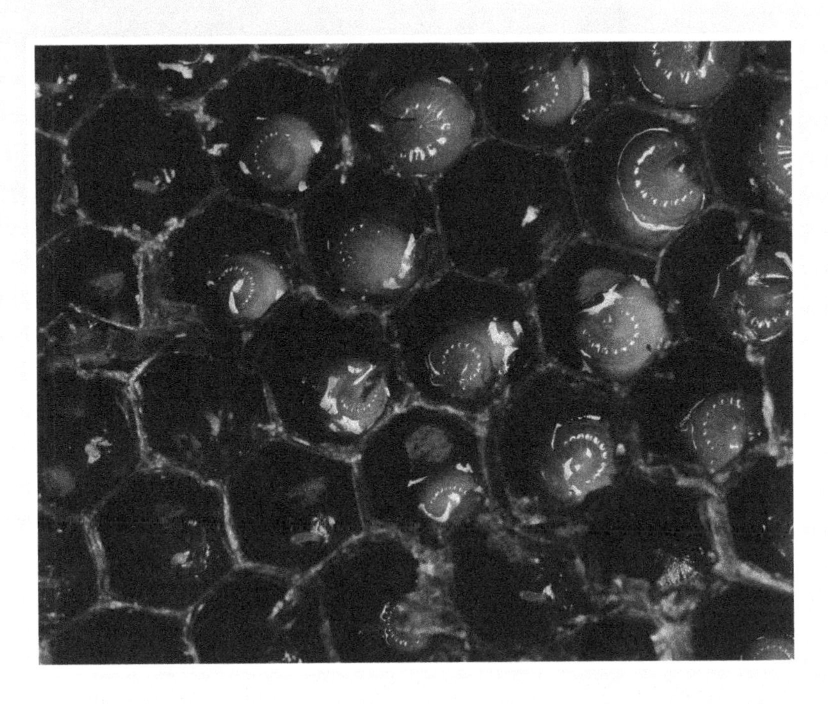

SWARMING

By now, you should have a good understanding of the honey beehive—what it looks like, the family members who live there, and how it functions. But what happens when a queen reaches old age?

At this point, the colony must divide through a process called *swarming*. Whereas a queen laying her eggs allows her to increase the population of her family, swarming allows for the colony as a whole to reproduce its social unit.

Swarming is an elaborate operation. Normal hive activities slow, sometimes to a full stop. Workers build special cups, hung vertically from the margin of the comb, and then direct or wait for the queen to lay fertilized eggs there. One of those eggs will become her replacement. She is not aware of the significance of this behavior—laying eggs in cups is merely a normal behavioral response to no longer being able to produce enough of her pheromone essence to satisfy her daughters and maintain her dominance.

Workers once so solicitous of their queen now put her on a crash diet; she will need to lose about half her body weight in nine days so she can fly away with the workers. Colony foraging decreases, and workers start to hang around with more honey than normal in their honey stomachs.

The departure usually occurs a day or so before the first newborn queen emerges. Bees begin to move rapidly about the hive, signaling that it's time to go. Weather permitting, the queen and about two-thirds of the adult workers, stomachs filled with honey, exit the hive, leaving behind the rest of the workers, all of their brood, their beeswax comb, and any honey they're unable to carry. They pour out the entrance and fly away in a circular mass, mindful of their queen. They land on a fence post, tree, or similar structure and form into a *swarm cluster* (see photo on page 90).

A swarm cluster is an awe-inspiring sight. Thousands of bees settle from the air, cling to one another, and settle into an oval mass, becoming quiescent in a matter of minutes. The *swarm site* is a temporary resting spot, a bivouac. The bees may stay clustered for only a matter of minutes, or they may stay for hours—even days. The only observable activity is scout bees flying from the cluster, dutifully searching cavities to find a suitable new home. Sometimes scouts find several options; they use dance language to "advertise" their cavity. A "debate" may break out. To determine which site is better, scouts will visit one another's sites. The best site will garner more excited dancers and win out. At that point, the entire cluster of bees becomes airborne again to fly directly to the chosen site, guided by scouts darting back and forth to keep the cluster together and headed in the right direction.

Meanwhile, back at the original hive, virgin queens emerge. Within a day, the hive settles on one, who will mate with several drones within a week or so. Now the parent hive—the hive that lost its queen with the swarm—has its new queen, capable of mating and laying eggs. The hive has successfully reproduced, or divided; where there was one, now there are two.

Beekeepers can intervene in hive-swarming behavior. When a swarm is spotted at the bivouac spot, they may seek to capture these free bees in an empty hive. Upon inspecting their hives, beekeepers may also discover developing queen cells and handle the division themselves, splitting the colony into two hives. However, swarming is the bee's natural method of reproduction, and pre-splitting the hive might not always be successful.

Bees in the Popular Imagination: Swarming and Democracy

Swarming—with its various candidates and heated debates and consensus-seeking cluster—may not be so dissimilar from democratic elections held in America. Professor Tom Seeley, of Cornell University, has devoted his life to studying how bees accomplish tasks in the hive. He had this to say about the process, in his excellent (and aptly named) book *Honey Bee Democracy*:

> *When a honeybee swarm chooses its future home, it practices the form of democracy known as direct democracy, in which the individuals within a community who choose to participate in its decision making do so personally rather than through representatives . . . It is a "unitary democracy" since it involves individuals who have congruent interests (choose the best home site) and shared preferences (small entrance opening, etc.) . . . [Their] debate works much like a [human] political election, for there are multiple candidates (nest sites), competing advertisements (waggle dances) for the different candidates, individuals who are committed to one or another candidate (scouts supporting a site), and a pool of neutral voters (scouts not yet committed to a site) . . . The election's outcome is biased strongly in favor of the best site because this site's supporters will produce the strongest dance advertisements and so will gain converts the most rapidly.*

Suffice it to say, a hive is an amazingly cohesive unit. As humans, we can't help but marvel at how well bees accomplish critical and everyday tasks.

CHAPTER 6

MODERN BACKYARD BEEKEEPING

This chapter provides an overview of modern beekeeping for hobbyists—the tools of the trade, the growing popularity of backyard beekeeping, best practices, and more. For those interested in hosting a colony of bees themselves, the essentials covered here might be a good jumping-off point; at the very least, I hope it opens up the possibility to take a more active role in saving the bees!

THE PERKS OF BACKYARD BEEKEEPING

People who love the outdoors, nature, animal care, and gardening also tend to love bees and beekeeping. This is not a coincidence. In addition to having a personal supply of fresh, process-free honey, backyard beekeeping can have "footprint" benefits for both the local environment and your community. Bees are the ultimate win-win-win—a win for you, a win for local wildlife, and a win for the environment.

When you start beekeeping, you unleash a small army of workers from your backyard who will go about pollinating flowers of all sizes, shapes, and colors within a range of two to three miles. Consider this a free service to your neighbors' gardens and the community. If you're an active gardener, you'll be doing yourself a favor, too—bees will boost your yield of fruit and/or vegetables. With pollination, bees will be helping flowers set seed to ensure diversity and color in your yard and beyond, which in turn helps feed and shelter local birds and wildlife.

Of course, on a more personal level, you will have some beeswax available, with which to make your own candles (see page 161), furniture polish (see page 166), lip balm (see page 165), and soap (see page 167). Propolis, easily collected from your beehive with a simple plastic hive insert, can be used as a skin wound dressing or digestive aid (see page 137). And by purchasing an inexpensive plastic pollen trap, you can trap pollen right at your hive entrance. It won't do any harm to your bees and may provide you with a range of health benefits (see page 138).

Certainly, to the uninitiated, starting your own backyard beekeeping operation might seem a bit intimidating. If you're serious about it, I recommend that you check out a more comprehensive, practical guide. (I've included a few of my favorite books and websites in the Resources appendix.)

A few words of wisdom: Have clear objectives and expectations. Don't fret over the quote-unquote "best" way to beekeep. There is no one correct way to keep bees. Keep an open mind, check several sources, and then make your decisions accordingly. Remember, you wanted to keep bees in the first place because it sounded like a fun and interesting activity—a challenge. As you embark on your beekeeping journey, don't let the obstacles you face get in the way of your enjoyment!

THE TOOLS OF THE TRADE

It may surprise you to learn that you won't need a ton of money to keep one or several beehives in your backyard. The cost of a startup colony ranges from $300 to $700 while adding additional colonies runs from $200 to $300.

Here's what you'll need to get started.

Protective Equipment

Beekeeping gives new meaning to the phrase "dress for success." What follows is a quick rundown of the most important gear necessary to avoid getting repeatedly stung.

Veil: You'll always want to wear a veil, even if you are approaching a hive for simple, quick tasks.

Clothing: You'll need to wear clothing that covers your entire body and periodically inspect bee clothing for tears or openings. One good option is to buy bee overalls that zipper tightly with a secure veil.

Footwear: Boots or work shoes are recommended when working with bees to protect your legs and ankles. Tuck your coveralls or pants into footwear or close pant legs/shirt cuffs with strapping to keep bees out.

Gloves: Gloves are a must when it comes to protecting your hands and wrists from stings. Tight-fitting gloves are best because they allow you to move nimbly within the hive and avoid crushing bees.

Bee Smoker: This is a pot that contains a smoky fuel with bellows to direct smoke into the colony, which changes guard bee behavior (see page 84) and allows beekeepers to more peacefully open the hive.

Hive Tool: This is a useful, spatula-like flattened tool to pry and separate the frames and assist in colony inspection with minimal disturbance to the bees.

THE BACKYARD BEEKEEPER'S GUIDE TO AVOIDING BEE STINGS

It's only natural for the first-time beekeeper to wonder: Now that I've decided to host a colony of bees in my backyard, how do I avoid getting stung? I've compiled seven rules to follow to protect yourself.

1. **Never open the bee colony without your personal protective equipment.** It is far easier to remove clothing, such as gloves, than to stop to put on such clothing after the bees begin to sting.

2. **Do not walk in front of the colony entrance.** Manipulate bee colonies from the side or back.

3. **Avoid alerting a colony with ground vibration disturbance,** such as weed trimmers or lawn mowers.

4. **Keep bee colonies on individual hive stands** to avoid having activity or vibrations pass from one colony to neighboring colonies.

5. **Be aware of bee behavior.** Manipulate potentially defensive bee colonies, those that too quickly react or take quickly to the air, with a helper or experienced mentor.

6. **Manipulate bee colonies only under favorable weather conditions.** Think 70°F or higher, little wind, sunny conditions and preferably midday (10 a.m. to 2 p.m.), when the majority of foragers will be outside their hive foraging. Keep it open only for a short time.

7. **Have a plan.** What would you do if you or someone else suddenly receives many stings? Have your cell phone on you in case of emergencies and keep topical sting remedies handy to help reduce the pain and discomfort, whether from a single sting or multiple.

Beehives

Each colony that you maintain will be housed in a hive. I recommend you start with the standard wooden Langstroth hive, available from bee supply dealers. For each box, you'll need 8 to 10 wooden or plastic frames with beeswax foundation, depending on the box size.

There are other hives you can readily purchase, such as *top bar* or *Warré*. These boxes do not have standard frames, simply a top bar. When bees build their combs, they do not necessarily align with your top bars and quickly become unmanageable (we call this "cross-comb"). As harvestable honey-filled combs are not extractable, the honey harvest is messy, requiring comb destruction. That said, after you master the movable frame hive, a top bar hive might be a fun (if occasionally challenging) alternative to try.

Feeders

Obviously, to start beekeeping you'll require bees (see page 100 for tips on acquiring them). And to feed your new colonies, you'll need sugar syrup, so start looking for granulated sugar sales. (Big box stores will have larger quantities at lower prices.) You can make feeders for the sugar syrup from recycled peanut butter jars or coffee cans with tiny holes punctured in the lids. Alternatively, you can purchase feeders—I recommend a refillable sugar water feeder so you can refill it without having to open the hive and disturb the bees. The best place to put feeders is at the top of the hive.

FAQ: Where Does One Actually Buy Bees?

Naturally, your hive is nothing without the bees. You might be able to go to your local hardware store and pick up a colony (if they handle bee supplies), but it's more likely you'll need to search for bee suppliers; many communities have a local beekeeper who also sells bee supplies and bees. Herewith, I've included several of the most common ways to secure bees for your backyard hive.

PURCHASE A NUC. A nuc (nucleus) colony is a smaller version of an established colony. Established beekeepers divide their strongest colonies, supply or let the division make a queen, and subsequently offer nucs for sale. Nucs are housed in a smaller moveable comb hive box, so they're readily transportable. And while the bees are established, nucs are smaller and far easier to open and inspect, though you'll need to transfer them to a standard hive eventually. This is a good option for beginners. You can gain confidence and knowledge as your colony grows.

ORDER PACKAGE BEES FOR DELIVERY. Incredible as it sounds, you can buy bees, like hamburger meat, by the pound. Three pounds fill a hive to start a new colony. The bees are shipped in a wooden or plastic box with screen siding. You can specify a delivery date—sometime in April typically works best for suppliers. Having a date in mind will give you the time (and motivation) to get you and your backyard prepared for the bees. You will have to install the bees in your hive. Make sure you read the instructions in the package itself—work up the courage, get a buddy to help, and at the end of delivery day, just do it! I promise you, it's not as difficult as you might think. Package bees are the preferred method for starting a colony.

BUY A COLONY FROM A FELLOW BEEKEEPER. If you check Craigslist or other such online classifieds, you may occasionally see BEEHIVE SALE listed. Prices will vary (anywhere from $100 to $250 or more). But beware, as occasionally only empty hive equipment is being offered. You will also need a plan to move it to your backyard. Plus, the question remains: Are you prepared to own and handle an established colony? Are you able to evaluate if you are buying a problem hive? The bees are in their established home, they know what they are doing, and they know how to protect their home. Are you prepared to jump right in? This option is often too challenging for a beginner.

CAPTURE A SWARM. If you hear of a clustered swarm at a nearby bivouac location (see page 91), you can capture a swarm by brushing or shaking them into your new hive equipment. Swarm bees are gentle because they tanked up before leaving home and are full of honey they've taken with them from the parent hive.

Capturing the swarm is technically the most cost-effective option—all you're spending is time—but it does not come without risk. Other beekeepers may beat you to it, the swarm may be clustered high up in a big tree or on someone else's property, or the bees may have left by the time you get there. Worse yet, the person reporting the swarm may confuse hornets for bees. You don't want to mess with hornets, so proceed with caution!

SITUATING YOUR HIVE

At one time, many farms had beehives. Some still do. Commercial beekeepers may or may not be farmers growing crops or keeping animals. With thousands of colonies, they don't own enough property, so they need to rent sites to park their beehives when the bees are not being used for pollination. For ages, bee colonies have typically been kept around the home residences where their keepers live.

Today, backyarders (or hobbyists) are the fastest-growing beekeeping group, as beekeeping-related courses fill to capacity with—if you'll forgive the pun—"newbees." They keep one, two, or even up to a half-dozen colonies, which are often considered pets. Usually, they aren't interested in maximizing honey production. Some are even bee-less, simply helping a friend or caring for a hive at a nature center or the local bee club. Hobbyists put their beehives in their backyards, on their porches, even on a city roof—anywhere out of the way (and out of their neighbors' way).

Personally, I would look for a lightly shaded spot with enough space for you to move around all sides of the hive. Don't fret if you can't find the perfect spot. If you have a storage shed (or the garage has become a storage shed), commandeer some space to store your protective equipment, tools, and hive parts.

Home beekeepers do have a responsibility to prevent their bees from visiting neighbor's pools, bird baths, or patios to collect their water. As such, you should locate a close watering spot. A container with rocks and molding leaves or a muddy spot nearby from a slowly dripping hose works well. An open container replenished with fresh water is less desirable, as bees like "smelly" water.

If you're a beginner, I recommend that you start with two hives. This will allow you to compare and contrast and interchange frames of honey or brood from stronger to weaker colonies.

URBAN BEEKEEPING

Simply put, urban beekeeping is keeping colonies of bees in urban areas. A widely shared BBC story recently claimed urban bees are "healthier and more productive than their country cousins"—though research, including a 2015 study from *PLOS One*, has suggested the exact opposite: Bees on rural farms and those properly cared for are the healthiest and most productive. However, proponents of urban beekeeping *do* correctly point out that urban beekeepers provide communities with environmental and economic benefits, such as plant pollination and locally produced honey.

Beehives on a rooftop garden in London.

At one time, cities discouraged beekeeping. Some had laws or ordinances that banned or restricted bees. But times and attitudes have changed. Today, many cities welcome, or at least do not discourage, beekeeping, including New York, Chicago, Detroit, Toronto, Vancouver, and San Francisco, just to name a few. When contested, continued restrictions simply cannot be justified. Local governments increasingly recognize the benefits of bees. Bans do nothing to reduce feral colonies in structures. And at the end of the day, bees are not a public health hazard. Though there is a dearth of statistics on the growth of urban beekeeping, it is evident that the practice is enjoying a renaissance, with beekeeping becoming more and more common in cities big and small, in community and rooftop gardens, as well as backyards.

Beekeepers in a number of states are required to register their hives, but enforcement is lax, and few urban beekeepers bother or are aware of registration requirements, especially in states that have a registration fee. To find out whether your community bans or restricts beekeeping, you have to ask. Sometimes local or state associations have (often incomplete) information on towns or cities that have restrictions in their area. Experienced beekeepers will advise you not to ask, recommending instead that you simply keep your hives hidden from prying eyes. Out of sight, out of mind.

However, virtually all communities have nuisance regulations, and if bees are kept in a negligent manner, you can be cited and asked to remove the bees. Interpretation of a nuisance may be exceedingly broad and/or a ruling of one public employee based on a complaint of a single neighbor. In these instances, solicit help from other beekeepers or your local bee association if you'd like to appeal. The best defense is to follow guidelines on best management practices (see the Resources appendix for more).

You Better Bee-Lieve It: High-Tech Advents in Beekeeping

Despite beekeeping's tactile pleasures, beekeepers are not immune to high-tech gadgetry and programs. Computer hardware, software, and related apps are now available to measure weather data, bee colony weight, and hive temperature; some can even detect special bee sounds, as in those used to fight robbers or in preparation to swarm. In the old days, you would have to ask your beekeeping friends in person or over the phone about the goings-on of their colonies. These days, you can use internet-based programs to add data from your hive into a shared database—not only do you get personal feedback, but you have access to data from other cooperators as well.

A quick internet search will provide you with a healthy list of companies and products associated with beekeeping supplies. Access to the technology and the technology itself are getting better and better all the time, but the devices can still be rather costly and may require specialized knowledge to be useful.

Of course, if you want to get in on the action without all the real-world cost (and work), places like Black Dogs Farm in Kansas are offering "virtual beekeeping": You can adopt and sponsor a hive, follow its progress online, and even keep a portion of the honey it produces—all from the comfort of your own home.

NATURAL BEEKEEPING

There are many opinions on what constitutes a *natural* approach to beekeeping. Natural beekeeping is more of a philosophy than an official set of rules. So, what principles should we include in this philosophy?

Ross Conrad, a Vermont beekeeper who has published the book *Natural Beekeeping, Organic Approaches to Modern Apiculture,* has developed a pretty solid concept of natural beekeeping. Conrad considers natural beekeeping as "honey bee stewardship that addresses pest, disease, and potential starvation issues without relying on synthetic pesticides, antibiotic drugs, or the regular use of an artificial diet . . . it does not necessarily mean minimal manipulations and it definitely does not mean minimal hive inspections . . . Minimal or no hive inspections is honey bee neglect, not natural bee stewardship."

Organic beekeeping is considered by some as the natural way. Unfortunately, there is no currently approved definition of organic beekeeping by the USDA. To be truly organic, you would need to be certified as following certain standards. Because your bees cannot be confined to property you personally control, such certification is hard to obtain, unless you live in an area where there are no agricultural or urban neighbors.

Large-scale beekeeping has many inherent risks for the beekeeper. Commercial beekeepers, along with virtually the rest of modern for-profit farmers, have elected to reduce potential risks with chemical pesticides, prophylactic antibiotic use, and supplemental dietary inputs. When keeping bees in heavily farmed regions, such externalities are considered necessary—just as necessary as they are for the neighboring farmers growing crops or raising livestock. Bees are their livelihood, and reducing risk-taking is part of farming, be it crop or animal. Some of these precautions are dictated to qualify for insurance or government subsidy payments, which are designed to help farmers continue to provide us with food at a relatively low cost.

Where does this leave you if you want to keep bees naturally? Your attitude toward prophylactic antibiotic use or how you feel about using a pesticide in or around your home may well come into play. Some of us immediately seek an antibiotic when we feel sick or a family member is ill. Others would prefer to wait it out and only take an antibiotic if absolutely necessary. Likewise, we all have varying attitudes on the use of pesticides. At the first sign of a bug, some reach for a pesticide to kill it. The only good bug, after all, is a dead bug! Others are willing to allow some pest and/or pathogen damage and only seek controls when absolutely necessary. You can be natural, following one or the other approaches.

Being a natural beekeeper takes extra effort. Natural means employing best management practices, controlling swarming and varroa mites, feeding bees when they need to be fed, and making sure to provide nearby sources of water. Rudimentary as they may seem, staying on top of these tasks will inherently reduce your dependence on chemical pesticides and antibiotics and help make natural beekeeping a viable option.

BEEKEEP RESPONSIBLY

Responsible beekeeping has several implications: It means being responsible to your bees by keeping them healthy and regularly inspecting for diseases, brood stress, or weak, unproductive colonies. It means being responsible to your neighbors by respecting their property and space. And it means being responsible to the local environment by employing sustainable and environmentally friendly beekeeping practices.

For urban beekeepers especially, responsible beekeeping requires paying attention to where the bees are placed and how and when they are inspected. By keeping bees, you can make a valuable contribution to your local ecosystem. But being a good neighbor is of the upmost importance. Opening your colony when the next-door neighbor is hosting a backyard barbecue party, for example, is not being a responsible beekeeper. Likewise, putting your beehive as far away as possible on your property, closer to your neighbor's property, is probably not the best call.

Bees often "out" the beekeeper when they go to the neighbors to collect water. Swimming pools have a chlorine smell that is attractive to bees, and bird baths are a convenient place for a bee to obtain water. Counter a bee's proclivity to search for water elsewhere by providing "dirty water" close to the colonies.

As we've discussed, a swarm can be a boon for starting a bee colony: It is not too hard to wrangle a cluster of free bees to start a new colony. But for non-beekeepers, swarms can incite fear. If a swarm leaves your backyard hive and lands on the fence or a tree on the neighbor's property, you are not being a responsible beekeeper. You might capture it as a service, but that alone might not be enough to mollify your neighbor.

In almost all these circumstances, the best practice is to stay ahead of potential problems. Some would advocate talking to neighbors before you even start your hive in order to mitigate future conflicts. However, if you are at odds with a neighbor, adding a beehive might merely be another point of contention.

With the prevalence of varroa mites (see page 36), sustainable and responsible beekeeping has become more and more challenging. Without pest control, bee colonies will die in two or three years—the varroa mite is that serious. Urban beekeepers frequently avoid using antibiotics or pesticides inside their bee colonies, sometimes simply from a lack of understanding. However, as a colony weakens from viral infections fostered by mites feeding on bee brood and adults, the sick colonies horizontally transfer mites—from one colony to another. In effect, the beekeeper who has elected to do nothing about mites is in fact condemning their neighbors' colonies and feral colonies to get sick.

The varroa mite also takes a toll on a colony's ability to perform the community service of plant pollination. Individual infested bees die early and do not forage as effectively or as long as healthy bees. Sick colonies do not produce sufficient honey stores for the beekeeper to harvest surplus. As a result, local nectar is not being collected and processed into honey for distribution among friends and neighbors.

Of course, some beekeepers prefer a natural or organic approach to beekeeping and mite control; although it's more challenging, this can work, too. Remember: There's no one right way to keep bees. But keeping bees healthy and productive while maintaining sustainable management practices is critical for all beekeepers, whether beekeeping for fun or for profit. Good beekeepers make for good neighbors, not only among rural folks but also in big cities.

PART III

The Bee Lover's Home and Garden

If you've made it this far, I'll assume you've come to not only love and admire bees but cherish their role in modern society. In this final section, you'll find some general tips for those looking to start and maintain a bee-friendly garden as well as an overview of honey, beeswax, and other bee products. The book will close with a selection of my favorite bee-centric recipes, from desserts like honey-drizzled baklava, to home goods like beeswax candles and soap.

CHAPTER 7

THE BEE-FRIENDLY GARDEN

The effort to save the bees can begin right at home, in your own garden. While the specifications of a bee-friendly garden will vary depending on where you live and how much space you can devote to it, the concept is the same: Plant bee-friendly plants, whether by layering flowering plants, shrubs, and trees or simply reducing the size of your lawn. Local bees will be better for it, as will your garden. What's good for bees is good for all of us.

CULTIVATING YOUR BEE-FRIENDLY GARDEN

As we discussed in chapter 1, the coevolution between plants and bees began in the Cretaceous Period (see page 15). That coevolution continues today. Suffice it to say: Bees and plants are as inextricable as plants and humans, bound together in a mutually beneficial (and ecologically significant) relationship. By proactively planting a variety of bee-friendly plants, we can provide a benefit to the bees, who can provide a range of benefits to the environment and those who live in it. Bee-friendly plants can make our world a brighter place, ushering other wildlife, such as birds and butterflies, into our everyday lives.

If you look online, you can find many helpful recommendations from a number of reputable organizations for how to become a bee-friendly gardener (see Resources appendix). In fact, it's relatively easy. And for most individuals who want to help bees, it is far more reasonable to cultivate a friendly bee habitat than to become a backyard beekeeper.

Read on for a general overview of a bee-friendly garden. And when you're ready to get started or modify your present garden, make sure to use the checklist at the end of this chapter.

On the Menu: Bee-Friendly Plants for Your Garden

You'll recall that bees, as strict vegetarians, rely on pollen for proteins, minerals, vitamins, cholesterol, plant phytochemicals—all the things needed for bees to reproduce and colonies to grow. To entice bees, plants offer sugar-rich, sweet-smelling nectar as a bribe, in expectation that the visitor will accidently transfer pollen from anther to stigma. Visiting flowers takes energy, and it's the nectar that supplies bees with their sugar. We can think of pollen providing bees with their *grow* power and nectar providing bees with their *go* power.

But there is more to the story. As explained in chapter 2, flowers developed specific modifications to attract specific visitors. With a few exceptions, neither flower nor pollinator bee is present the entire season. These specific modifications, coupled with specific pollinators, enabled flowering plants to spread and inhabit all types of habitats.

In your own garden, you might consider plants that attract the particular bees you prefer. Some plants, such as those with long, tubular flowers, like lilac, primarily attract bumble bees; mason bees, on the other hand, like flowers buried within vegetation, such as ivy, and ground-nesting bees like smaller, native flowers. Generalist bees, like honey bees, seek open flowers with a strong fragrance; as such, some flowers attract only a few bees while others attract a wider variety. If you're observant, you might even see some friendly competition as different bees jockey for position on a flower.

As growing conditions change from season to season, so too will the flowers. That's why cultivating a variegated garden is appealing to bees. Some flowers are more cold-hardy, appearing when temperatures are still cold at night and the sun is low on the horizon. Early bloomers like crocus, calendula, snowdrops, borage, mustards, and dandelion and trees and shrubs such as winter jasmine, winter honeysuckle, alder, manzanita, pussy willow, Pieris (Japanese andromeda), red maple, and red bud can help provide the pollen that bees need to begin their reproduction cycle.

During the spring, as weather warms, a wider variety of plants and bees appears. Anemone, clematis, native lilies, dahlias, peony, lilac, phlox, poppies, zinnia and wild roses, and shrubs and trees such as apricot, orange, viburnums, rhododendron, and azalea—all of them really shine during the spring. Yet it's important to remember that some plants, like forsythia, double cherries, or designer roses and other intensely modified, showy flowers, though they may be beautiful and long-lasting, may in actuality have little or nothing to offer bees. When in doubt, do the shake and sniff test. Brush the flower in your hand and sniff the bloom; if you see no pollen in your hand or there is no odor, then it is unlikely to attract bees.

During the summer, many of the same plants listed previously continue in bloom. Shaded areas, irrigation (in areas lacking natural rainfall), and planting space will greatly influence your plant selections. Generally, extending the spring into summer bloom will take more effort on your part. Some great summer flowers include bee balm, cosmos, echinacea, snapdragons, foxglove, and hosta; flowering trees like mimosa, chestnut, paulownia, sourwood, American fringe tree, golden chain tree, and Chinese flame tree; and shrubs such as cotoneaster, hollies, ivy, lilac, clethra, hibiscus, and hydrangea. Not all varieties are necessarily practical for all gardens, and I have not sorted out weedy species or specified native versus non-native—that will be for you to decide.

In fall, garden plants often get shortchanged. We tend to simply run out of energy, the weather gets hot, or life otherwise gets in the way. Plus, our selections get scarcer. The great fall landscape plants of the east—aster, smartweed, ironweed, and goldenrod—are less common in the west. Desert areas become very dry and hot. Some autumnal garden bloomers include zinnias, sedum, calendula, and witch hazel. That said, you can always rely on virtually any of the herbs, like rosemary, mints, lemon balm, and echium, and weeds such as dandelion and clovers to prosper. Some of these re-bloom with cooling temperatures and fall precipitation. We can seek a continuous bloom with supplemental watering and proper planting sites.

For more specific information on attracting native bees, I recommend checking Xerces Society's *Attracting Native Pollinators*. Naturally, though, the makeup of your garden is subject to your own whims, limitations, and climate. Part of the fun comes from making your own selections.

You Better Bee-Lieve It: The Rules of Attraction

Flowers use a variety of "weapons" to attract bees; chief among them are color and smell. Scientists believe foraging bees are initially attracted to color, as they generally fly about a foot above the vegetation, while scent helps them home in on odors appealing to them—particularly sweet, floral scents. Consequently, flowers planted in groups will be more attractive than a single flower.

Flowers, of course, come in a kaleidoscope of colors. But bees do not perceive color the same way humans do. Whereas our base colors are red, blue, and green, bee base colors are blue, green, and ultraviolet light, which our eyes cannot see. Subsequently, where we see red, bees see black, making red flowers much less appealing to them. As the renowned Dr. Karl von Frisch observed, in general, bees are most attracted to blues, whites, pinks, and yellows in pastel hue; they're much less drawn to red, dark blue, and dark purple.

While flower petals may appear as a single color to our eyes, they may have *nectar guides*—lines of ultraviolet color extending out from the center, creating a "bull's-eye" effect. Such lines guide bees to the center of the flower.

Secondarily, the prominent position of flowers and a flat, open structure both add to its allure. Because bee eyes react better to moving objects than stationary ones, they may also be seduced by the way a flower dances in the breeze. A grouping of plants that all flower at the same time in a single area, due in part to their sunlight, soil, and moisture requirements, presents a super stimulus. Due to flower constancy—the tendency of bees to forage only one kind of flower at one time—foragers will still only select one flower type among a panoply of flower colors in simultaneous bloom. Not that they're picky, but bees are good at learning which flowers offer the rewards they seek.

Interestingly enough, there are a few bees that are attracted to flowers for their oils; most of these bees live in tropical regions.

The Go-To Plants for Every Region

If you've had an opportunity to travel this country, you've surely noticed that the vegetation varies from region to region, from palm trees in Florida, to an abundance of sweet clover in the Dakotas, to miles of sage brush in the Southwest, to spruce and pine in the Pacific Northwest. As you look around the suburbs, however, especially where water is supplied, lawns will often look quite alike, and yards will often have similar flowers, shrubs, and trees.

To become a good bee-conscious gardener, it would be helpful to consult the USDP plant hardiness zone map, also termed a planting or growth zone map (see the Resources appendix). In total there are 13 zones, with Zone 1 being the northernmost, and Zone 13 being the southernmost. (Zones 5, 6, and 7 stretch from New England to California, including more than 50 percent of the regions where Americans live.)

Each zone averages 10 degrees warmer or colder in the winter than the adjacent zone; zones where most of us live are subdivided. For example, in Pennsylvania there are three zones, each subdivided. The map is interactive, so you can look up your specific zone and use it to help determine which plants might be best for where you garden. Ultimately, the zones help identify plants that are adapted to the winter cold, summer heat, and temperatures in between.

A number of resources list appropriate plants for each of the zones. Nurseries often specially grow plants suitable for a particular zone—though not all of these are bee-friendly. If you live in an urban area, this might not be as cut-and-dried—asphalt and buildings become heat sinks and radiate heat after the sun sets, creating hotter (and drier) conditions than might be indicated for your zone. Consequently, you might consider taking a soil sample to your cooperative extension for testing. This will help determine if you need to augment the soil or add fertilizer for what you are planning to grow.

When all is said and done, all gardening is local. If you're in doubt, it's best to seek advice from local nurseries and garden specialists.

Going the Extra Mile: Additional Accessories for Your Bee-Friendly Garden

Before you get started, take a gander at this list of supplementary bee-centric accessories.

Bee hotels/condos are a combination of nesting attractions, which may include hollow stems, blocks of drilled cut wood pieces, and stones, in between which bumble bees can find hospitable cavities. Packing straw into a sheltered structure is also appealing, as bees like to rest (and nest) in straw. In general, such additions make for practical and interesting garden "sculptures," sure to rouse bee and human curiosity alike.

Bee houses are structures mason bees need for nesting—essentially, wooden blocks with drilled holes or straws with one open end. Though they're sold commercially, they could also make for a fun and easy DIY project (see photo on page 119).

Bee spas. Every good hotel needs a spa, right? Honey bees require water for hive air-conditioning when it gets hot or to dilute honey. They will overlook clean water for "dirty" water, such as wet soil areas, containers with decaying leaves/vegetation, and water with chlorine (i.e., swimming pools). Including a bee spa involving plants and perching areas will provide that necessary water.

Seed bombs. Planting seed bombs is sometimes referred to as "guerilla gardening," because you're planting seeds outside your property. You can make seed bombs by mixing one handful of bee-friendly plant seed to three handfuls of potter's clay to five handfuls of peat-free compost, adding enough water to form them into hand-size balls. Allow them to bake in sun for three hours before distributing. Be responsible, though: Vacant lots, overgrown weedy areas, and otherwise unused areas are okay, but avoid other people's yards and landscaped public areas. (See the Resources appendix for more details.)

Yard signs. Various programs, including Audubon and Pollinator Protection, sell yard signs for bee lovers to help spread the word. They demonstrate your commitment and are a great conversation starter to educate inquisitive neighbors who ask about what you're doing.

The Bee-Friendly Gardener's Checklist

From growing the right plants to avoiding pesticides, use this checklist, compiled from a number of sources, including World Bee Day, Friends of the Earth, Garden Know How, and Gardenista, to make sure your garden is bee-friendly.

❑ **PLANT FOR POLLINATORS**

Not all flowers offer bees comparable rewards. In general, you'll want to choose long-blooming plants that have colorful, aromatic flowers. (Pass by the showy flowers, hybrids, or double flower varieties, especially if they lack fragrance.) Look for plants with high numbers of flowers, even if flower size is small. Plant in groups, as large as your space will allow. Keep an eye on what's blooming in your garden and make notes on where you see bees. When in doubt, it's a good idea to check with your local nursery, as many now put a bee-friendly label on their plants. Find websites for flowering plants appropriate for your plant hardiness zone or ask the master gardener specialists at your local cooperative extension office (each county has one). Research what they advise and ask follow-up questions.

❑ **PLANT FOR THE SEASONS**

When does your garden lack flowers? Bees need flowers for their total adult life. Use the On the Menu section (see page 113), research the entire flower season, and seek to fill in any gaps when there are fewer flowering plants in your garden.

- [ ] **THINK LIKE A BEE**

  What do bees need? Food, yes, but also shelter. If you love bumble bees, provide overwintered queens a place to nest. Mining bees need dry soils, so make a berm and don't overwater it. Grow or distribute hollow reeds, such as phragmites or bamboo, or place drinking straws (paper, please!) and/or wooden blocks drilled with holes in shelters for mason bee nest sites. (Alternatively, you might purchase commercial mason bee houses and even buy some bees to stock them.)

- [ ] **WATER, WATER, WATER**

  Your plants need it, of course, but so do bees. That's why you should include some type of water feature in your garden. Bees like water with a smell, so incorporate water plants and let decaying vegetation accumulate. Bees also need to perch, so you'll want to include stones or plant features they can use (as opposed to simply a big pool). And be sure to plan some time to sit and watch your water collectors! Likely, you'll see more at the beginning and the end of each day.

- [ ] **ALLOW NATURE TO BE MESSY**

  For the benefit of the bees, you'll want to weed less, allowing the bees' favorite flowers to dominate the weeds rather than the other way around. Lawns with weeds provide foraging opportunities, while carpets of a single grass variety are deserts and will take considerable gardening time to maintain. Reduce lawn size and plant larger borders in vegetative layers (with the lowest adjacent to the lawn, increasing in height in subsequent layers). Small and medium-size trees can offer great numbers of flowers. Three layers are quote-unquote "easy"; try for four or five.

- [ ] **CUT OUT THE CHEMICALS**
Minimize your use of insect-killing pesticides and plant-killing herbicides. Don't ignore pest outbreaks, but seek to spot-treat early before pest populations build—and only when necessary. Like humans, unhealthy plants don't look as well as healthy ones, and when they're ailing, they may have little to offer bees. For proper maintenance, seek alternatives to hard pesticides, such as organic chemicals or eco-pesticides, like soaps or oils. And don't be afraid to welcome and foster beneficial creatures, including frogs, ladybugs, lacewings, and other wildlife, to your yard.

- [ ] **CONSIDER LANDSCAPE VARIETY**
Ground-nesting bees need well-drained soils, such as a berm. Piles of rocks or stacked wood are great potential nest sites. Leave areas less heavily mulched so ground-nesting bees can better access the soil surface. If you see ground-nesting bees, seek to protect or expand that area. It has something they need, and continued maintenance will be easier as a result.

- [ ] **THINK LONG-TERM**
Make this a fun project. Purchase something to transplant immediately, and then make a wish list of what to follow up with. Seeds will take longer to produce flowers, but there's joy in watching the plants themselves develop. Start small, but start now.

- [ ] **SPREAD THE WORD**
Individual actions multiplied across a community increase the speed of progress. Spreading the word yields many rewards benefiting the whole community. Sometimes all that's needed is the spark. Be that spark! It takes a community to effect real change, but it all starts with passionate individuals.

CHAPTER 8

HONEY, BEESWAX, AND OTHER BEE PRODUCTS

Ah, honey: a term of endearment for a special person as well as the sweet, natural product harvested from honey bee colonies. Although honey is by far the most familiar honey bee product, it is actually but one of several unique materials obtained from bee colonies. Besides collecting flower nectar to convert into honey, honey bees collect pollen and propolis and produce beeswax, royal jelly, and bee venom, all of which can prove useful to humans. In this chapter, we'll dive headlong into the wondrous bounty of bees.

THE TRUTH ABOUT HONEY

Though everyone generally knows what honey is and where it comes from, not everyone knows how, exactly, it's created. Honey is actually a processed food, though it's processed exclusively by honey bees. Bees collect nectar from flowers and add the enzyme *sucrase* from their salivary glands, which reduces the complex plant sugars of nectar into two primary sugars: glucose and fructose. These sugars are immediately absorbable by bee and human digestive systems.

When bees return with nectar, they spread many droplets across the hive, speeding up the evaporation process; with excess water removed and a high concentration of converted sugars, the nectar—now honey—both conserves storage space in the hive and doesn't spoil with time. The sweetness results from the highly concentrated sugars. But honey is more than its sugars. The particular amino acids, enzymes, minerals, phytochemicals, and numerous other plant components unique to the nectar source are still present. These are what provide the distinctive flavors and colors of honey.

So, how does honey get from the bee to our table? It starts with beekeeper harvest of filled honeycombs. Beekeepers may cut the comb and sell fully ripened comb honey just as the bees stored it. Following harvest, the bees need to construct replacement combs. More commonly, though, beekeepers extract the honey using centrifugal motion in an extractor (quite literally spinning it out of the comb) and then let it settle, filtering out air bubbles and pieces of beeswax. The harvested combs are left intact and returned, minus the honey, to the hive for refilling.

The liquid, also called extracted honey, is packaged in bulk or bottled in containers of various sizes and shapes; plastic bears are popular, as is a special flattened honey bottle design designed to show off the honey's color. By controlling crystal size, beekeepers can solidify the liquid to produce creamed honey—a smooth, butter-like honey spread.

You Better Bee-Lieve It: The Magical Properties of Honey

As the reverence for honey bees worldwide shows, honey is something of a magical ingredient. Everyday people use it directly as a sweetener on toast and in tea and as a sugar substitute in a wide variety of foods. Athletes like it during high-intensity competitions because no digestion of the sugars is needed. Bakers like to use it to keep their baked goods fresher because honey, unlike sugar, absorbs moisture from the air.

In addition, much of the world uses honey as a medicine. Honey can soothe sore throats and ear aches, relieve allergy symptoms, and serve as an antibacterial antiseptic (skin burns, wounds, and/or lesions) and as a diluent to help the medicine go down. It can also be a great hangover remedy—although perhaps not after a night of drinking mead (see page 157)!

As a food, honey offers special dietary advantages. Because it's made of two simple sugars, it can be absorbed from our intestinal tract directly into our bloodstream—no additional digestion required. Consequently, honey can replenish blood sugar levels quickly. That's why you see a number of energy drinks using honey as a sweetener. People on diets also like honey because honey packs more sweetening power than the same amount of sugar.

As Joe Traynor documents in his book *Honey: The Gourmet Medicine*, humans have used honey as a medicine for as long as we've been using it as a sweetener. Fully ripened honey produces small amounts of hydrogen peroxide, making it a mild diuretic and antiseptic for the skin, both human and animal. One particular honey type from New Zealand, known as Manuka, can withstand pasteurization—which is used to get rid of potentially harmful microorganisms—without losing its antimicrobial properties, so it's widely accepted for wound dressings. That said, virtually any honey is effective, and the chances of bacterial infection are very rare, as honey viscosity is not conducive to survival or growth of harmful microorganisms.

Generally, wounds heal faster with less scarring when honey is used. Honey may also be an effective treatment for certain infections that have developed resistance to antibiotics, such as hospital staph and methicillin-resistant Staphylococcus aureus (MRSA). In his popular book *The Honey Revolution,* Dr. Ron Fessenden recommends a teaspoon of honey before bed to restock the liver with glycogen and slightly raise insulin, improving the brain's ability to function and one's ability to sleep through the night.

Kirsten S. Traynor reviews medical information related to honey, much of it published outside the United States. In her book *Two Million Blossoms: Discovering the Medicinal Benefits of Honey,* Traynor describes why honey helps heal chronic wounds, how it bests antibiotic-resistant superbugs and reduces tissue scarring and nerve damage, and even its potential to improve memory. She also notes the ways in which it can minimize the debilitating side effects of cancer treatments. According to Traynor, some in the medical community are more receptive to trying honey therapies when conventional treatments fail to provide relief (and as we continue to learn about the dangers of multiple drug combinations).

Regardless of whether you're planning on using honey to help alleviate plant-based allergies, sore throats, upset stomachs, or other medical or dietary issues, you'll want to look for local honey (sometimes labeled "raw" or "natural"). But beware: While there are specific requirements necessary to label "organic" honey, no such requirements exist for use of the words "raw" or "natural." When in doubt, ask local honey producers—they'll be delighted to share with you where their honey came from and how it was produced.

A BEE LOVER'S GUIDE TO BUYING HONEY

Like wine or coffee, a honey's particular character is in part the result of the unique climate, soil, and moisture conditions of the flower from which bees collect nectar. Just as sommeliers can determine a wine's region from its taste, so, too, can experienced apiarists identify a honey's floral origin based on its unique color, flavor, and aroma. And there can be a world of difference. Honey high in fructose will be sweeter, while others

will taste fruity or earthy, and darker honey has a stronger taste. As you might with wine or unfamiliar foods, taste first with your nose, then swirl the initial spoonful thoroughly around your mouth and breathe through your nose after swallowing. Savor it all.

Using dance language (see page 87), bees concentrate their foraging to flowers with higher sugar content. Some distinctive uni-floral honeys include basswood honey from upper New York State, clover honey from the Dakotas, orange blossom honey from Florida, sage honey from California, and sourwood honey from the Carolinas and Georgia. Cultivated crop plants, such as the sunflower from the Midwest, lima bean from Delaware, or cotton from Texas, also produce distinctive honeys. When bees visit multiple flower types, the result is multifloral wildflower honey, which will vary from region to region and season to season.

Here are a few frequently asked questions about honey, answered. (For more information, I recommend consulting the National Honey Board website.)

What is the difference between raw or natural honey and processed or filtered honey?

When beekeepers extract honey, air bubbles, pieces of wax, and pollen may get mixed into the liquid. Raw or natural honey means the liquid is minimally processed, with just the larger particles filtered out. When such liquid honey crystallizes (solidifies), the sugar crystals form around these particles and crystals become larger, making the honey's texture gritty and unappealing.

To give the liquid honey a longer shelf life, honey is heated and forced through a series of fine-mesh filters, thereby delaying crystallization. To provide a consistent pack throughout the year, this filtered honey is frequently blended with honey from national and international sources. In both instances, the honey is still pure.

Is raw or natural honey better for you than processed or filtered honey?

The National Honey Board's 2012 annual report analyzed vitamins, minerals, and antioxidant levels in raw and processed honey. The study showed that processing/filtering significantly reduced the pollen content of the honey but did not affect the honey's nutrient content or antioxidant activity.

If there's no difference between raw and filtered honey, why might I want to buy local raw honey?
Blended, processed honey is mainly a dietary sweetener—the "generic" honey, if you will. Natural or raw honey identified by floral source or harvested from a specific location offers distinctive tastes and colors; it's also considered the superior source in matters of medicine. Plus, the purchase supports local beekeepers.

What is organic honey?
According to the USDA, organic honey must be harvested from bee colonies whose management adheres to organic livestock standards. Standards state that the hives and the flowers from which the bees will be getting their nectar must be free of chemicals. Once removed from the hive, the honey can be strained but not heated.

However, since beekeepers cannot control where their bees forage—a two-mile radius from the hive—virtually no US honey qualifies for an organic label.

How are various kinds of honey categorized?
Honey may be categorized by floral source or location where produced, but it is more frequently categorized by color (light, amber, or dark) and USDA grade. When harvested in the comb, it is sold as comb or section honey. When beekeepers cut the comb to fit packaging, it's called *cut-comb honey*. When they spin the liquid honey from the wax comb, it is termed *extracted honey*, or simply liquid honey. Liquid honey will eventually crystalize—this does not mean it's spoiled, merely solidified.

Where are the best places to purchase honey?
As you well know, you can purchase processed, blended honey in supermarkets or most food outlets. Local varietal or artisanal honey is sold at farm stands and farmers markets as well as at specialty food stores or on the internet. Such honeys will be unique to location or floral source. Check the website of your state beekeeper's association for a listing of beekeepers selling local honey; many associations have a buyers' directory.

How should I store my honey?
Store honey at room temperature. If you purchase a large quantity at one time, store it in the freezer until ready to use to preserve color and flavor; freezer storage is the best way to delay crystallization.

My jar of honey has turned from liquid to solid. Is it still good? Should I throw it out?
Your honey has simply crystallized! To reliquefy, remove the lid and put the jar into a warm water bath or microwave until it turns back into a liquid. Be careful to not overheat it, as that will degrade the color and flavor. After it reliquefies, it will likely recrystallize more rapidly than it did the first time.

Is there an expiration date on honey?
Nope! While honey may darken over time, a closed container is self-preserving due to the production of a small amount of hydrogen peroxide and high viscosity (low moisture). If germs get into an open container, they simply are unable to grow in this environment. So, rest assured, honey is perfectly edible even after years of storage; unlike wine or cheese, the taste does not perceptibly change. Here's a little-known fact: The oldest known honey, which dates back 5,500 years ago, was discovered in the tomb of a noblewoman in Georgia, not far from Tbilisi, still perfectly good to eat. And the honey buried in jars in King

Tut's tomb? It was still delicious and edible 3,000 thousand years later!

How do I properly substitute honey for sugar in a recipe?
Seeing as honey has superior sweetening power, you can use ½ to ⅔ cup of honey for every cup of sugar called for in a recipe. The National Honey Board also recommends that for each cup of honey used, you subtract ¼ cup of other liquids and add ¼ teaspoon of baking soda. If you're making this substitution for a baked good, then lower the oven temperature by 25 degrees.

You Better Bee-Lieve It: Bee Venom Therapy

The last time you (or someone you know) was stung by a bee, you likely noticed, besides the initial few minutes of pain, a persistent swelling and irritation, which may have lingered for four or five days. Essentially, the bees that stung you wanted you to avoid bothering them a second time. These normal reactions are due to complex chemicals in the venom that affect normal body cellular functions. And though you may personally think its powers are being used for ill, bee venom can also be used for good!

As Kirsten S. Traynor details in *Two Million Blossoms*, bee venom has long been recognized for its medicinal properties. For humans suffering from conditions such as rheumatoid arthritis, nerve pain (neuralgia), multiple sclerosis (MS), swollen tendons (tendonitis), and certain muscle conditions, such as inflammation of muscles and/or joints, bee-venom therapy has been reported to ease pain and suffering. That said, the venom is not a cure, nor will it halt further deterioration of these conditions. But it will provide temporary relief.

Bee venom is collected by shocking bees on an electric grid, causing them to sting through a special thin plastic sheet. The venom droplets are frozen and, in a sterile environment, scraped from the plastic sheet and processed. Medical doctors (typically allergy specialists) are the only individuals to prescribe bee sting immunotherapy, via injection of select chemical components, void of the irritating and pain-inducing properties. There are, however, unlicensed "practitioners" who apply actual bee stings directly to exposed skin numbed with ice. (Unfortunately, the pain and irritation components remain intact.)

BEES, WHAT ARE THEY GOOD FOR? (NOT JUST HONEY)

Although it was honey that first drew humans to bees, we've since grown to harvest a range of additional products from our most important insect ally. On top of their invaluable pollination services, bees provide a veritable treasure chest of riches for us to use, as outlined in the pages to follow.

Beeswax

Pound for pound, beeswax, a complex compound of fatty acids and various long-chain alcohols, is worth more than honey. Worker bees produce this wax in glands of the abdomen. Bees mold the wax into parallel beeswax combs and use it to cap cells of fully ripened honey (see page 83).

HOW WE HARVEST IT

Honey cell cappings provide the most desirable, lemony-colored beeswax. For every 100 pounds of honey harvested, beekeepers obtain one to two pounds of wax. In a *destruct harvest*, whereby honey-filled combs are crushed and the liquid honey drained, wax yield will be higher. Additionally, beekeepers may accumulate bits and pieces of wax and any pieces of extra comb removed during colony inspections. Beeswax crafters can purchase beeswax directly from a beekeeper or from craft marketers.

WHAT WE USE IT FOR

Three major products, in more or less equal amounts, consume the annual harvest of beeswax: candles, cosmetics (like lipsticks, lotions, and bath products), and beeswax foundation (see page 66).

Beeswax is also used as:

- Lubrication for needles and nails as well as an industrial lubricant
- Dental impression wax
- Pill coating
- Floor, furniture, or lens wax
- Musical instrument finish
- Insulation for electrical wiring
- A preventive for saltwater corrosion
- Molding for jewelry and figurines
- Ski, snowboard, and surfboard wax
- Sealant for food containers and wine bottles

Propolis

Propolis, sometimes called bee glue, is the sticky, resinous, gum-like substance that some plants secrete to cover flower buds and the flow of sap from certain trees. Bees mix this resin with saliva to strengthen beeswax comb and reduce comb vibration. But that's not all. Bees also use it as an antimicrobial coating for frames and boxes, to close off spaces smaller than bee space (see page 65), and to smooth rough surfaces in the beehive. Additionally, bees use propolis to entomb dead hive intruder bodies (like mice), coat the entry and exit areas of the nest, and reduce hive draftiness during the fall and winter.

The color and chemical composition of propolis are pretty complex and vary greatly by plant source. In only a few instances has propolis been characterized chemically; it is mainly a mixture of plant resins, oils, wax, and, in minor amounts, amino acids, minerals, sugars, vitamins, flavonoids, phenols, and aromatic compounds. Propolis is sticky at higher temperatures greater than 68°F while hard and brittle at lower temperatures.

How We Harvest It

Propolis can be scraped from bee boxes and frames, but such scrapings may include beeswax and other impurities. It is far easier to use a propolis trap, a plastic grid with tiny oval spaces placed at the top of a colony, beneath the inner cover. Bees will fill in the tiny spaces with propolis. The plastic sheet can be put in freezer, and, once frozen, you can pop out the pellets, like ice cubes from an ice tray. Another method is to create finger-size openings in the side of the bee box. Bees will fill the space with propolis, which can then be scraped into a container. Both latter methods yield high-quality, uncontaminated propolis.

To extract the chemicals from harvested propolis, soak it in alcohol for six to nine months (agitating occasionally). Gentle heating speeds the processing; boiling off excess alcohol gives you the final liquid. For human consumption, the final liquid tincture is adjusted for about a 15 percent concentration; for use on our skin, it may be more concentrated.

Given the complexity of this process, your best source of propolis is the beekeeper collecting it.

WHAT WE USE IT FOR

Traditionally, propolis has been used as a human medicine, although modern Western medicine considers its efficacy unsubstantiated. That said, homeopathic and Eastern medicines have used it as a covering for wounds and cuts; in liquid form, it's been taken to help digestive disorders and as a daily dose for promoting digestive regularity. Non-medically, it's been used as a varnish, polish, and protective coating; it is widely used for wooden string instruments, like the violin. A number of commercial finishing varnishes include propolis.

Pollen

Pollen is the male germ cell of the plant and the major food for bees, supplying them with amino acids, proteins, lipids (fats), minerals, vitamins, plant phytochemicals, cholesterol—in short, everything bees need to grow.

HOW WE HARVEST IT

Beekeepers collect pollen via an entrance trap that removes pollen basket loads from the hind legs of returning foragers. To help it adhere and stay within the pollen baskets, bees mix nectar and saliva with the pollen grains themselves. It appears as pellets in the pollen trap, which are then winnowed to remove bee legs and impurities before being air-dried. You can purchase pollen directly from beekeepers who harvest it; it's also widely available in nutritional outlets, packaged in bottles or capsules for human consumption.

WHAT WE USE IT FOR

Pollen contains vitamins, minerals, and antioxidants, making it the darling of health-conscious individuals; however, affirmative medical studies are lacking. Among claimed benefits of pollen are decreased inflammation, improved immunity, relief from menopausal symptoms (such as hot flashes), and wound-healing promotion. Individuals with some plant-based allergies should get tested for possible allergic reactions before consumption.

Pollen may be fed directly to, or mixed into various high-protein feeds for, the bees themselves, as well as feeds for pet fish, reptiles, and birds.

CHAPTER 9

BEE-CENTRIC DRINKS, RECIPES, AND CRAFTS

In the previous chapter, we went over the kinds of products we can derive from bees and what we can use them for. In this chapter, I'll provide you with some practical, fun, and easy ways to use the marvelous products harvested from our bees, including food and drink recipes, as well as clear instructions on how to make a few home goods, from beeswax candles to soap. Take these as a jumping-off point: Try them out, play around with them, and make sure to share with your family, friends, and neighbors. Soon enough, they'll be just as passionate as you are about enjoying the harvests from bees!

Classic Honey Toddy

SERVES 1 / **PREP TIME:** 3 TO 5 MINUTES

Ideal for winter parties, having the gang over for cards, or as a nightcap on a cold evening, the secret to a good hot toddy is the right mix of lemon and honey. If you're suffering from a cough or sore throat, go heavier on the lemon and a little lighter on the whiskey; if you don't like it too sweet, cut back a bit on the honey.

¾ cup water

1½ ounces whiskey of your choice

2 to 3 teaspoons lemon juice (preferably freshly squeezed)

2 to 3 teaspoons local raw or natural honey

1 cinnamon or peppermint stick, for garnish (optional)

1. Pour the water into a kettle and bring it to boil over high heat; then remove from the heat.
2. Pour the hot water into a mug or glass with a handle. Then add the whiskey and lemon juice.
3. Stir in the honey and insert the cinnamon stick (if using). Serve hot.

HONEY TIP: **Get creative and try honey from different floral sources—you will be surprised how a honey's floral source can reinvent the traditional toddy.**

Cool Honey Lemonade

MAKES 2 QUARTS (OR 8 [8-OUNCE] GLASSES) / **PREP TIME:** 10 MINUTES, PLUS 1 HOUR TO MARINATE AND 1 HOUR TO CHILL

The balance of sweet and sour flavors in this refreshing beverage will have your guests coming back for more during your next warm-weather gathering. For a nice twist, add fresh strawberries or cranberries when serving; feel free to use a mix of lemons and limes. If your first batch is too sweet, cut back on the amount of honey.

---

8 lemons (or mix lemons and limes)

8 to 10 teaspoons honey

2 quarts cold water

Ice

Lemon slice and/or mint leaves (optional)

1. Thinly grate the lemon skins into a bowl. Mix in the honey and let stand for 1 hour.
2. Squeeze the lemons into another bowl and add the cold water. Combine with the honey-marinated lemon skins and place in the refrigerator until cold, 30 to 60 minutes.
3. Filter the lemon skins. Add ice and garnish with a lemon slice and/or mint leaves (if using) on the glass rim.

HONEY TIP: **Darker-colored honeys can be stronger tasting and produce a darker, less appealing lemony color. Ideally, you should be able to taste both the honey and lemon in your drink.**

Norma's Cereal Bars

MAKES 20 TO 30 BARS / **PREP TIME:** 30 MINUTES, PLUS 24 TO 48 HOURS TO DRY

Norma is a queen-bee breeder I work with on bee programs in Bolivia. She has two growing sons who come home from school famished. She makes these delicious cereal bars for them, but never fails to bring me some as well. Fortunately, I managed to get the recipe out of her to share with you.

5 cups oatmeal

1 cup seeds (chia, sesame, pumpkin, and/or sunflower)

1 cup dried shelled nuts (walnuts, almonds, and/or peanuts)

1 cup dehydrated fruit (raisins, cranberries, plums, bananas, and/or pineapples)

1 cup honey

Vegetable oil cooking spray

1. In a medium frying pan, toast the oatmeal over medium heat, about 5 minutes. Remove from the pan and repeat this process with the seeds and shelled nuts.
2. In a bowl, thoroughly mix the oatmeal, seeds, shelled nuts, and dehydrated fruit with the honey, stirring vigorously for 10 to 15 minutes.
3. Spray a flat tray with vegetable oil cooking spray and spread the mixture on top. Flatten with a rolling pin or clean bottle to make sure the mixture is about an inch thick and flat on top.
4. Let dry, uncovered, on the counter for 24 to 48 hours, or until the ingredients are completely clumped together.
5. When dry, spray a pizza cutter or sharp knife with vegetable oil and cut the mixture into bars (about 1 by 3 inches).
6. If you're saving for later, tightly wrap each bar in plastic wrap.

Amish Blueberry Cornbread

MAKES 9 SQUARES / **PREP TIME:** 10 MINUTES / **COOK TIME:** 30 MINUTES, PLUS 10 MINUTES TO COOL

I used to frequent a Pennsylvania Amish farmers market to buy whoopee pies. On one visit, they didn't have any left, so the sales clerk suggested I try blueberry cornbread instead. It soon became a household favorite. I kept returning to ask about ingredients and cooking methods, and after some trial and error, I was finally able to make my own. For best results, eat it fresh out of the oven—with honey drizzled on the edges.

2 large eggs

¾ cup buttermilk

¼ cup honey, plus more for drizzling

¼ cup butter, melted

⅔ cup all-purpose flour

1½ cups cornmeal (if less cornbread flavor is preferred, adjust with flour)

1½ teaspoon baking powder

½ teaspoon salt

1 cup fresh or frozen rinsed blueberries

1. Preheat the oven to 375°F.
2. In a medium bowl, whisk together the eggs, buttermilk, honey, and butter.
3. In another medium bowl, mix together the flour, cornmeal, baking powder, and salt.
4. Stir the dry mixture into the egg mixture until evenly moistened, being careful not to overmix. Gently fold in the blueberries.
5. In a buttered 8-inch square pan, spread the mixture evenly with a spoon or spatula.

CONTINUED >

Amish Blueberry Cornbread CONTINUED

6. Bake for 25 to 30 minutes, or until a wooden toothpick inserted in the center comes out clean.

7. Let cool for 10 minutes; then cut into 9 squares. Drizzle the cut edges with honey before serving.

HONEY TIP: **For the final honey drizzle, you'll want a lighter, milder-flavored honey so as not to overpower the cornmeal and blueberry taste.**

Traditional Baklava

SERVES 8 TO 12 / **PREP TIME:** 30 MINUTES / **COOK TIME:** 45 MINUTES, PLUS SEVERAL HOURS TO COOL

Honey is hygroscopic, meaning it absorbs moisture from the air. This helps keep your baked goods moister and fresher for a longer period of time. Baklava is a traditional honey-sweetened, buttery dessert, typically found in Greece and areas throughout the Middle East. To prepare it requires a bit of attention to detail. But all your work will be worth it once you start getting rave reviews from friends and family. For the nuts, try using a mixture of pecans and pistachios.

---

2 (1-pound) packages frozen phyllo dough

¾ cup butter, divided

4 cups chopped nuts

1 teaspoon cinnamon

⅛ teaspoon nutmeg

2 cups honey

½ cup sugar

1 tablespoon vanilla extract

½ cup water

1. Thaw the frozen phyllo dough overnight in the refrigerator and remove 1 hour before using. Be sure to remove only the sheets you need at the moment, keeping the other sheets covered in plastic wrap and a damp cloth.
2. Preheat the oven to 350°F.
3. In a small saucepan or microwave, melt ¼ cup of butter.
4. In a small bowl, combine the nuts, nutmeg, and cinnamon.
5. Brush a rectangular baking pan with the melted butter. Check to see if the sheets of phyllo will fit comfortably in the pan. If they won't, trim them to fit or use a larger pan.

CONTINUED >

Traditional Baklava CONTINUED

6. Using the same melted butter from the pan, butter the top sheet of phyllo. Then set this top sheet and the unbuttered sheet directly below it into the pan, buttered-side down. Press them lightly into the pan.

7. Repeat step 6 two more times, so that you have 6 sheets of phyllo in the pan, half of which are buttered.

8. Sprinkle on a heaping layer of the nut mixture. Butter 2 sheets of phyllo and place them on top of the nuts, buttered-sides down. Repeat this process, adding layers of nuts topped with 2 buttered phyllo sheets until you're out of the nut mixture. Then top with 4 more buttered phyllo sheets, buttered-sides down.

9. Cut the baklava into diamond-like shapes using a very sharp knife. Bake until golden brown, about 45 minutes.

10. While the baklava is baking, combine the honey, sugar, vanilla, water, and remaining ½ cup of butter in a saucepan. Bring the mixture to a boil, stirring vigorously; then reduce the heat to low.

11. Remove the baklava from the oven. While it's still hot, drizzle half of the honey-vanilla mixture evenly over the top. Allow it to sit and absorb for a minute; then drizzle on a little more, until the surface is thoroughly moistened. You may not need (or want) to use all of the honey-vanilla mixture.

12. Allow the baklava to cool, uncovered, for several hours before serving.

13. Reheat in the microwave for about 10 seconds so it's nice and hot before serving.

HONEY TIP: **You want to select a lighter-colored honey with a fruity or earthy taste for this recipe.**

Honey Peanut Butter Cookies

MAKES 10 TO 16 COOKIES / **PREP TIME:** 10 MINUTES / **COOK TIME:** 15 MINUTES

Who can resist peanut butter cookies? Others sweeteners, like sugar or even maple syrup, may be used here, but I, of course, prefer to stick with honey.

1 large egg, beaten

1 cup creamy peanut butter

¼ cup oat bran

1 teaspoon vanilla extract

¼ cup honey

Sea salt, for sprinkling

1. Preheat the oven to 350°F.
2. In a large bowl, combine the egg, peanut butter, oat bran, vanilla extract, and honey and mix with a spatula until well combined.
3. Roll level tablespoons of the batter into balls. Place on an ungreased baking sheet and flatten the balls with a fork.
4. Bake for 15 minutes, until golden brown in color. Let cool on a wire rack and sprinkle with sea salt before serving.

Anna's Peanut Butter Nuggets

MAKES 6 TO 8 BITE-SIZE NUGGETS / **PREP TIME:** 5 MINUTES

This is an ancient recipe, according to my beekeeper friend Anna. As a kid, it was her favorite after-school snack for her and her siblings. For extra credit, you can make your own peanut butter. (When I do, I like to add a little honey when I'm blending my peanuts with peanut oil.)

1 tablespoon honey

2 tablespoons powdered milk

2 tablespoons peanut butter

Finely chopped peanuts, for coating (optional)

1. In a medium bowl, put the honey. Mix in the powdered milk, followed by the peanut butter.
2. Shape the mixture into bite-size balls and enjoy. You can roll the balls in finely chopped peanuts (if using) as a surface coating. Or, if you're really hungry, you can eat them directly from the bowl!

HONEY TIP: **Any honey will do here, as you are seeking that mid-afternoon energy boost; the peanut butter will largely overwhelm the honey flavors.**

Chef Steve's All-Purpose Meat Glaze

MAKES ABOUT 1 CUP (OR ENOUGH FOR ABOUT 2 POUNDS OF MEAT) / **PREP TIME:** 5 MINUTES

Chef Steve Cohen's meat glaze is especially suited for game meat, such as venison, elk, wild boar, lamb, and the like. It has a tendency to lessen the gaminess and enhance the natural flavor of the meat. Personally, I like it as a ham glaze, though it goes well with beef, too.

¼ cup honey

½ cup brown sugar

¼ cup low-sodium soy sauce

¼ cup cheap bourbon

1 tablespoon balsamic vinegar

1 tablespoon Dijon mustard

1. With a spoon or spatula, hand-blend the honey, brown sugar, soy sauce, bourbon, balsamic vinegar, and mustard, making sure the brown sugar has been completely combined. (You can reverse portions of brown sugar with honey if a slightly sweeter glaze is desired.)
2. Depending upon the meat you intend to glaze, the glaze can be injected with a cooking syringe or brushed on the surface. If it's the latter, replenish periodically as the meat cooks.
3. Before serving, add the final surface glaze as the meat cools.

HONEY TIP: **Honey works great as a meat glaze either by itself or as a carrier for other spices and flavors you wish to include (as with this recipe). In general, honey provides a nice counterbalance to acidity and spice. I use it in chili—and even spaghetti!**

Sweet and Salty BLT

MAKES 4 SANDWICHES / **PREP TIME:** 10 MINUTES / **COOK TIME:** 15 MINUTES

Introducing the bee lover's version of the BLT: crispy, honey-brushed bacon topped with lettuce, tomatoes, and a honey mustard spread, all served on a salt-speckled pretzel bun.

4 tablespoons honey, divided

1 teaspoon hot water

12 strips bacon

¼ cup Greek yogurt

2 tablespoons Dijon mustard

4 pretzel rolls

4 to 8 leaves butter or romaine lettuce, divided among 4 sandwiches

1 to 2 medium-size tomatoes, sliced and divided among 4 sandwiches

Sweet butter pickles (for serving)

1. Preheat the oven to 350°F.
2. In a small bowl, mix 2 tablespoons of the honey with the hot water until combined.
3. On a baking sheet, lay the bacon strips in one layer and bake for 10 minutes. Pat with a paper towel to soak up excess fat. Brush the bacon with the honey mixture. Then bake for 5 to 6 minutes more, or until crisp.
4. In a small bowl, mix together the Greek yogurt, mustard, and remaining 2 tablespoons of honey.
5. Generously spread the mustard mixture on both sides of each pretzel roll.
6. Place 1 or 2 lettuce leaves on the bottom of each roll, top with 3 strips of bacon and the tomato, and close with the top of the roll. Serve with sweet butter pickles.

Honey-Soy Bee Brood

MAKES 1 CUP / **PREP TIME:** 15 MINUTES / **COOK TIME:** 10 MINUTES

In the past, I served this as a final treat after a beekeeping course. This will be a hit for some and a miss for others. If you've never tried brood before—and many haven't!—it tastes like fried nuts or pork crackling. Add your favorite hot sauce or salsa for a different flavor.

1 cup drone pupae removed from a bee colony

3 tablespoons honey

3 tablespoons soy sauce

Hot sauce or salsa

Cooking oil

TO REMOVE THE DRONE PUPAE

1. Remove the brood frame with the capped drone pupae. (The drone capped cells will be rounded and stick up above comb surface.)
2. Using a serrated bread knife, remove the cappings from the drone pupae and shake the frame onto a plate or pan with a downward snap. Pat dry.

TO MAKE THE BROOD

3. In a medium pan greased with your favorite cooking oil, fry the pupae over medium heat for 3 to 5 minutes, until golden brown.
4. In a small bowl, mix together the honey and soy sauce. Add the mixture to the pan and stir until the pupae are completely coated, 3 to 5 minutes.
5. Add hot sauce or salsa to taste and serve hot.

Dewey's Mead

MAKES 8 TO 10 (750ML) WINE BOTTLES / **PREP TIME:** 1 HOUR, PLUS 2 TO 3 DAYS TO STERILIZE / **FERMENTATION PERIOD:** AT LEAST 7 WEEKS / **STORAGE:** STORE IN CORKED WINE BOTTLES IN COOL, DARK SITE

Mead is thought to be one of the oldest fermented beverages—if not the oldest. It's made from diluted honey sugars, fermented by yeast to produce an alcoholic beverage, with anywhere from 3 to 15 percent ABV depending on the amount of sugar and yeast. Mead may be prepared exclusively with sugars of honey. Alternatively, honey may be mixed with other fruit sugars or herbs, spices, or virtually anything else—even your hottest habanero peppers!

Like most beverages, meads will vary depending on the maker's personal taste. Some like a very sweet wine, made with ferments that start with a higher amount of honey (not my preference). If you like semisweet or dry wines, you'd want to lower the sugar content. Regardless, you should plan on experimenting to hit on your particular preferences. Like all wines, mead will need some aging.

Here is a starter recipe for a semisweet mead; for more recipes, check out Ken Scramm's excellent encyclopedia of mead The Compleat Meadmaker *or* Making Mead (Honey Wine) *by Roger A. Morse, my mead mentor.*

TOOLS

2 (2½-gallon) plastic carboy containers

Hydrometer

Fermentation lock

Plastic tubing (to siphon liquid off dead yeast)

Wine bottles, corks, and bottle corker

CONTINUED >

Dewey's Mead CONTINUED

INGREDIENTS

7 pounds mild-flavored honey, plus more as needed

2 gallons chlorine-free water, plus more as needed

3 Campden tablets

2¼ teaspoons cultured wine yeast (high-quality champagne or madera-type wine yeast)

3 teaspoons citric acid

¾ pint strong tea

1 (2-ounce) packet yeast nutrient powder

Whites of 1 egg, for fining (optional)

1. In a carboy container, dilute the honey with water. This mixture is termed "must." Reserve a 2- to 3-ounce portion to start the cultured yeast. Then add the sterilizing Campden tablets to the rest of the mixture, and let it sit uncovered for 2 to 3 days.
2. Adjust the total sugar content so that it's anywhere from 22 percent (dry or less sweet) to 25 percent (sweeter) by adding up to 3 pounds more honey and up to three pints more water (keep in mind, you still need to add the tea). Using a hydrometer, the measured specific gravity should be 1.095 (dry) to 1.110 (sweeter).
3. Add the cultured yeast and cultured wine yeast to the honey-and-water mixture. Then add the yeast nutrient powder blend, citric acid, and tea.
4. Stopper the container with the fermentation lock to keep out wild yeasts and bacteria.
5. Leave the must to ferment in an out-of-the-way place. (Ideally, the temperature should be consistent, around 65°F.) Fermentation will begin as the yeast feeds on the diluted honey and releases carbon dioxide. If fermentation does not start within a couple of days (no bubbling), swirl gently and move to a

slightly warmer place. Check periodically over the first 3 weeks to ensure that the fermentation lock is still in place.

6. After 3 weeks, bubbling should be slow, and a whitish, dead yeast layer will be evident on the bottom of the container. Using the plastic tubing, siphon the liquid off this layer of dead yeast into another clean container and replace the fermentation lock. Be careful not to shake or mix the very bottom layer with the liquid above. Leave the siphoned mixture to continue fermenting for one more month. Discard the bottom mixture.

7. Next, siphon the liquid a second time into a clean container. If the liquid is cloudy after the second siphoning, you can precipitate the protein (i.e., the reason for the cloudiness) by adding the egg whites (a fining agent). The suspended protein will make your wine less attractive but otherwise not affect the taste.

8. You are now ready to sample your mead. With two hands, swirl a small amount in a wine glass, sniff deeply, sip, and rinse it around your mouth, breathing back through your nose. You can swallow (or, if the taste is immediately unpleasant, spit it out). Breathe back through your nose again. Your wine will be immature and likely have an acidic or alcoholic taste at this time. If it has a strong vinegary taste, your fermentation lock failed and bacteria turned the alcohol into acid. If your wine has qualities you like, such as a pleasant aroma, and is good enough to swallow and merits another taste or two, then bottle it (with a cork, not a screw top) and allow it to age.

9. The yield here should be 8 to 10 bottles. Take one out every few months or so to see if they're ready or if further aging is needed. (With this recipe, I've aged mine for up to several years.)

Beeswax Candle

YIELDS 16 FLUID OUNCES OF BEESWAX/ **PREP TIME:** 1 HOUR, PLUS UP TO 1 DAY TO HARDEN

The secret to great, sweet-smelling beeswax candles is using clean beeswax and the right wick—otherwise, candles may spit and give off black smoke. One way to clean the final impurities in beeswax is to strain melted beeswax through nylon stockings. Beeswax candles can burn hot, so mixing with a softer wax will bring down the melting point and allow a more even burn. Of course, you don't want it too soft, or your candle will burn away too quickly.

If you follow these instructions, you will yield approximately 16 fluid ounces of melted beeswax (so if the volume of your mold is 4 ounces, you can make 4 candles). You can adjust the amounts as needed, keeping the 1:1 ratio the same. The wick size will depend on the dimensions of your candle mold. The rule of thumb, however, is to select a wick size that would sufficiently melt half of the finished candle. For example, if your candle diameter is 1 inch, you would use a wick that is suitable for a ½-inch burn.

And remember: It is better to have leftover melted beeswax for next time rather than not having enough to begin with.

TOOLS

Double boiler, for melting wax

Immersion thermometer

Candle molds (polyurethane, plastic, glass, or aluminum)

Appropriate wicking for your mold (cored wicks are recommended for larger diameter candles)

Wire or paperclip

Modeling clay (if you plan on reusing mold)

CONTINUED >

Beeswax Candle CONTINUED

INGREDIENTS

1 pound beeswax, thoroughly cleaned

½ cup coconut oil

1. Using a double boiler, melt the beeswax over low to medium heat, gently stirring occasionally. (Do not mix air bubbles into the melting wax.) Do not excessively heat the water—keep it below boiling point. Use an immersion thermometer to monitor the melted wax's temperature—it should melt around 145°F to 150°F; do not let the temperature exceed 170°F, as the wax will lose its distinctiveness. This will take up to 30 minutes, depending on the water temperature and amount of wax you are melting. When wax is thoroughly melted, gently stir in the oil.
2. Prepare the molds with wicking by centering the wicking at the open top with a wire or paper clip. Leave the wick tail (to start burn) at the bottom. (You can always cut it later so it's the desired length.)
3. If your mold has been used before, the wick opening can become enlarged and leak wax. Seal the wick opening from the outside with modeling clay after wicking, if necessary. If you are using a glass jar or cylinder for your mold, center the wick as securely as possible.
4. Carefully pour melted wax into the mold from the open bottom. (If using a glass jar, you will fill it from the top.) If the mold is supersize, top the pour after several minutes. Beeswax tends to crack, so place the poured mold in a warm place, like the oven on its lowest temperature setting. Turn it off after a few minutes, but leave the solidifying candle in place.

5. Repeat with additional molds, if using.

6. When each candle is completely solidified (depending upon the size, this may take more than a day), do a test burn for at least 1 hour to determine if it's burning evenly or only in the center, if the burn is too rapid, or if there is black smoke or spitting and sparking. If the quality of the burn is not as anticipated, it likely means that the wax was not thoroughly cleaned in advance or that you need to alter the size of wicking.

Peppermint Lip Balm

YIELDS 9 FLUID OUNCES / **PREP TIME:** 1 HOUR, PLUS 5 TO 6 HOURS TO COOL

A beeswax lip balm moisturizes naturally, smells pleasant, and lets you kick the chemicals of commercial lip balms. Using more beeswax will make a firmer final product, while using a bit more shea butter will make it softer. This yield should fill approximately 60 (0.15-ounce) lipstick tubes.

TOOLS

Plastic lipstick tubes and/or lip balm tins

Plastic pipette or eyedropper, for filling tubes

Double boiler

INGREDIENTS

3 ounces beeswax, thoroughly cleaned

3 ounces coconut oil

3 ounces shea butter

1 to 2 dozen peppermint essential oil drops

1. In a double boiler, melt the beeswax, shea butter, and coconut oil over low to medium heat, stirring occasionally, 20 to 30 minutes.
2. Remove the double boiler from heat, but keep the melted ingredients in the double boiler so the mixture remains melted.
3. Add the essential oil a few drops at a time; then smell to make sure the scent is to your liking.
4. Using a pipette or eyedropper, quickly fill the tubes or tins with beeswax mixture. Do not let the mixture cool too rapidly or allow air pockets to form in the tubes.
5. Cool at room temperature for 5 to 6 hours before capping the tubes or tins. If the beeswax is separating from the tubes' sides, place them in a warmer temperature setting, like the oven on its lowest temperature, and allow to cool more slowly.

Beeswax Furniture Polish

MAKES ABOUT 3 TO 4 CUPS / **PREP TIME:** 45 MINUTES, PLUS SEVERAL HOURS TO COOL

Take some beeswax, add some mineral oil, drop in some lemony essential oils, and you've got yourself an all-natural, fresh-scented furniture polish, perfect for the wooden surfaces in your home.

⅔ cup beeswax, thoroughly cleaned

3 cups mineral oil

30 drops grapefruit seed extract or vitamin E

10 drops lemon essential oil

1. In a double boiler, heat the beeswax and mineral oil over medium heat for about 30 minutes, or until the beeswax is fully melted, stirring occasionally.
2. Add the grapefruit seed extract and essential oil and stir well. Pour the hot mixture into a dry container. Avoid air pockets by pouring evenly and slowly.
3. Let the mixture cool into a semi-hard consistency. This may take several hours.
4. To use, dip a rag into the polish and work it into wooden surfaces in a circular fashion. Let stand for a couple hours; then buff to a sheen.

Shea Butter and Beeswax Soap

MAKES 6 (2.4-OUNCE) SILICONE SOAP BARS / **PREP TIME:** 10 MINUTES, PLUS SEVERAL HOURS TO 1 DAY TO HARDEN

Here's an old family recipe for soap with a small yield to try. After this basic recipe, go ahead and try to develop your own distinctive soap products with other fragrances or additions.

7 tablespoons olive oil

4 tablespoons lemony-yellow beeswax, thoroughly cleaned

7 tablespoons shea butter

Essential oil of your preference (optional)

1. In a microwave-safe container, combine the olive oil and beeswax.
2. Microwave in 30-second bursts, until the beeswax is completely melted. Usually, 3 to 4 times is sufficient.
3. Remove the bowl from the microwave using oven mitts, as the mixture will be very hot.
4. Stir in the shea butter.
5. Stir in the essential oil (if using) a few drops at a time, until the desired fragrance is achieved.
6. Carefully pour the mixture into each of 6 soap bar molds.
7. Allow the mixture to cool and harden for several hours to a full day.
8. Store in a cool, dry place out of the sun to prevent melting.

Measurement Conversions

VOLUME EQUIVALENTS (LIQUID)

US STANDARD	US STANDARD (OUNCES)	METRIC (APPROXIMATE)
2 tablespoons	1 fl. oz.	30 mL
¼ cup	2 fl. oz.	60 mL
½ cup	4 fl. oz.	120 mL
1 cup	8 fl. oz.	240 mL
1 ½ cups	12 fl. oz.	355 mL
2 cups or 1 pint	16 fl. oz.	475 mL
4 cups or 1 quart	32 fl. oz.	1 L
1 gallon	128 fl. oz.	4 L

VOLUME EQUIVALENTS (DRY)

US STANDARD	METRIC (APPROXIMATE)
⅛ teaspoon	0.5 mL
¼ teaspoon	1 mL
½ teaspoon	2 mL
¾ teaspoon	4 mL
1 teaspoon	5 mL
1 tablespoon	15 mL
¼ cup	59 mL
⅓ cup	79 mL
½ cup	118 mL
⅔ cup	156 mL
¾ cup	177 mL
1 cup	235 mL
2 cups or 1 pint	475 mL
3 cups	700 mL
4 cups or 1 quart	1 L

OVEN TEMPERATURES

FAHRENHEIT	CELSIUS (APPROXIMATE)
250°F	120°C
300°F	150°C
325°F	165°C
350°F	180°C
375°F	190°C
400°F	200°C
425°F	220°C
450°F	230°C

WEIGHT EQUIVALENTS

US STANDARD	METRIC (APPROXIMATE)
½ ounce	15 g
1 ounce	30 g
2 ounces	60 g
4 ounces	115 g
8 ounces	225 g
12 ounces	340 g
16 ounces or 1 pound	455 g

Resources

BOOKS

100 Plants to Feed the Bees by the Xerces Society

Attracting Native Pollinators by the Xerces Society

The Backyard Beekeeper: An Absolute Beginner's Guide to Keeping Bees in Your Yard and Garden (4th edition) by Kim Flottum

The Beekeeper's Handbook (4th edition) by Diana Sammataro and Alphonso Avitabile

Beekeeping for Dummies (4th edition) by Howland Blackiston

The Bees in Your Backyard by Joseph Wilson and Olivia Messenger Carril

Beeswax Crafting by Robert Berthold, Jr.

The Compleat Meadmaker: Home Production of Honey Wine from Your First Batch to Award-Winning Fruit and Herb Variations by Ken Schramm

A Field Guide to Honey Bees and Their Maladies by Maryann Frazier, Dewey M. Caron, and Dennis vanEngelsdorp

The Hive and the Honey Bee by L. L. Langstroth (edited by Joe Graham)

Honey Bee Biology and Beekeeping by Dewey M. Caron and Larry J. Connor

The Honey Connoisseur: Selecting, Tasting, and Pairing Honey, with a Guide to More Than 30 Varietals by Marina Marchese and Kim Flottum

Making Mead (Honey Wine): History, Recipes, Methods and Equipment by Roger A. Morse

Storey's Guide to Keeping Honey Bees: Honey Production, Pollination, Health (2nd edition) by Malcolm T. Sanford and Richard E. Bonney

Why Do Bees Buzz?: Fascinating Answers to Questions about Bees by Elizabeth Capaldi Evans and Carol A. Butler

INTERNET

Bee City USA (for Bee Campus information, see page 48)
BeeCityUSA.org

Black Dogs Farms (for virtual beekeeping, see page 105)
BlackDogsFarm.net/products/virtual-beekeeping

Gardening Know How
GardeningKnowHow.com

Gardenista (for making your own seed bombs, see page 118)
Gardenista.com/posts/diy-wildflower-seed-bombs

High Country Gardens (for finding bee-friendly plants)
HighCountryGardens.com/plant-finder/bee-friendly-plants

Honey Bee Health Coalition
HoneyBeeHealthCoalition.org

National Honey Board
Honey.com

Oregon State University Best Practice Guidelines for Nuisance-Free Beekeeping
Catalog.Extension.OregonState.edu/em9186

USDA Plant Hardiness Zone Map
PlantHardiness.ars.usda.gov/PHZMWeb

US Forest Service (for more information on pollinators)
Fs.fed.us/wildflowers/pollinators

References

CHAPTER 1

Capinera, John L., ed. "Family: Andrenidae." In *the Encyclopedia of Entomology*, 429. New York: Springer, 2008.

Engel, M. S., I. Hinojosa-Diaz, and A. P. Rasnitsyn. "A Honey Bee from the Miocene of Nevada and the Biogeography of *Apis* (Hymenoptera: Apidae: Apini)." *Proceedings of the California Academy of Sciences* 60, no. 3 (May 7, 2009): 23–38.

Exley, Elizabeth M. "New Species and Records of *Quasihesma* Exley (Hymenoptera: Apoidea: Euryglossinae)." *Australian Journal of Entomology* 19, no. 3 (September 1980): 161–70. doi.org/10.1111/j.1440-6055.1980.tb02082.x.

Goulson, Dave. *A Sting in the Tale: My Adventures with Bumblebees*. London: Penguin Books, 2014.

Poinar, G. O., Jr., and B. N. Danforth. "A Fossil Bee from Early Cretaceous Burmese Amber." *Science* 314, no. 5799 (October 2006): 614. doi:10.1126/science.1134103.

Main, Douglas. "World's Largest Bee, Once Presumed Extinct, Filmed Alive in the Wild." *National Geographic* (October 2019). NationalGeographic.com/animals/2019/02/worlds-largest-bee-rediscovered-not-extinct.

The Hebrew University of Jerusalem. "First Beehives in Ancient Near East Discovered." ScienceDaily (September 5, 2007). ScienceDaily.com/releases/2007/09/070904114558.htm.

CHAPTER 2

Alaux, Cedric, Yves Le Conte, and Axel Decourtye. "Pitting Wild Bees against Managed Honey Bees in Their Native Range, a Losing Strategy for the Conservation of Honey Bee Biodiversity." *Frontiers in Ecology and Evolution* 7, no. 60 (March 13, 2019). doi.org/10.3389/fevo.2019.00060.

Almond Board of California. *2018 Annual Report.*. Newsroom.Almonds.com/sites/default/files/pdf_file/Almond_Almanac_2018_F_revised_4.pdf.

Calderone, Nicholas, W. "Insect Pollinated Crops, Insect Pollinators, and US Agriculture: Trend Analysis of Aggregate Data for the Period 1992–2009."

PLoS One 7, no. 5 (May 22, 2012): e37235. doi.org/10.1371/journal.pone .0037235.

Caron, Dewey M., and L. J. Connor. *Honey Bee Biology and Beekeeping*, 2nd ed. Kalamazoo, MI: Wicwas Press, 2017.

Delaplane, K. S. and D. F. Mayer. *Crop Pollination by Bees*. UK: CABI Publishing, 2000.

Food and Agriculture Organization of the United Nations. *Why Bees Matter: The Importance of Bees and Other Pollinators for Food and Agriculture*. 2018. FAO.org/3/I9527EN/i9527en.PDF.

Goodrich, Brittney. "A Bee Economist Explains Honey Bees' Vital Role in Growing Tasty Almonds." The Conversation. August 17, 2018. TheConversation.com/a-bee-economist-explains-honey-bees-vital-role -in-growing-tasty-almonds-101421.

National Research Council. *Status of Pollinators in North America*. Washington, DC: The National Academies Press, 2007. doi .org/10.17226/11761.

USDA Releases Results of New Survey on Honey Bee Colony Health." *U.S. Department of Agriculture*, May 12, 2016. National Agriculture Statistics Service. https://www.usda.gov/media/press-releases/2016/05/12/usda-releases-results-new-survey-honey-bee-colony-health.

US Geological Survey. "Why Are Bees Important?" Accessed April 1, 2020. USGS.gov/faqs/why-are-bees-important?qt-news_science _products=0#qt-news_science_products. (Also referenced in Introduction.)

CHAPTER 3

Aizen, Marcelo A., and Lawrence D. Harder. "The Global Stock of Domesticated Honey Bees Is Growing Slower Than Agricultural Demand for Pollination." *Current Biology* 19, no. 11 (2009): 915–18. https://doi.org/10.1016/j.cub.2009.03.071.

Bartomeus, Ignasi, John S. Ascher, Jason Gibbs, Bryan N. Danforth, David L. Wagner, Shannon M. Hedtke, and Rachael Winfree. "Historical Changes in Northeastern US Bee Pollinators Related to Shared Ecological Traits." *Proceedings of the National Academy of Sciences* 110, no. 12 (April 2013): 4656–60. doi.org/10.1073/pnas.1218503110.

Bee Informed Partnership. *Loss & Management Survey*. Accessed March 31, 2020. BeeInformed.org/citizen-science/loss-and-management-survey.

Berenbaum, May. "Bot-Flying." *American Entomologist* 65, no. 2 (June 7, 2019): 76–78. https://doi.org/10.1093/ae/tmz029.

"Bumble Bee Conservation | Xerces Society." Xerces Society for Invertebrate Conservation. Accessed May 19, 2020. http://xerces.org/bumblebees.

Ellwood, Elizabeth R., Stanley A. Temple, Richard B. Primack, Nina L. Bradley, and Charles C. Davis. "Record-Breaking Early Flowering in the Eastern United States." *PLoS ONE* 8, no. 1 (January 16, 2013): e53788. doi.org/10.1371/journal.pone.0053788.

Hopwood, Jennifer, Aimee Code, Mace Vaughan, David Biddinger, Matthew Shepard, Scott Hoffman Black, Eric Lee-Mäder, and Celeste Mazzacano. *How Neonicotinoids Can Kill Bees: The Science behind the Role These Insecticides Play in Harming Bees*, 2nd ed. Portland, OR: Xerces Society for Invertebrate Conservation, 2016.

PNW Honey Bee Survey Results. Accessed March 31, 2020. PNWHoneyBeeSurvey.com/survey-results.

Ramsey, Samuel D., Ronald Ochoa, Gary Bauchan, Connor Gulbronson, Joseph D. Mowery, Allen Cohen, David Lim, et al. "*Varroa Destructor* Feeds Primarily on Honey Bee Fat Body Tissue and Not Hemolymph." *Proceedings of the National Academy of Sciences* 116, no. 5 (January 29, 2019): 1792–1801. doi.org/10.1073/pnas.1818371116.

Soroye, Peter, Tim Newbold, and Jeremy Kerr. "Climate Change Contributes to Widespread Declines among Bumble Bees across Continents." *Science* 367, no. 6478 (February 2020): 685–88. doi.org/10.1126/science.aax8591.

US Department of Agriculture. Honey Bee Colonies (2019). https://usda.library.cornell.edu/concern/publications/rn301137d.

Watson, Kelly, and J. Anthony Stallins. "Honey Bees and Colony Collapse Disorder: A Pluralistic Reframing." *Geography Compass* 10, no. 5 (May 6, 2016): 222–36. doi.org/10.1111/gec3.12266.

CHAPTER 4

Ahnert, Petra. *Beeswax Alchemy: How to Make Your Own Soap, Candles, Balms, Creams, and Salves from the Hive.* Bloomington, IN: Quarry Books, 2015.

Arista Bee Research Foundation. "Aristaeus, the Greek God of Bee-Keeping." Accessed March 30, 2020. AristaBeeResearch.org/about-arista.

Diamond, Jared. *Guns, Germs, and Steel: The Fate of Human Societies.* New York: W. W. Norton & Company, 1999.

Herring, Peg. "Dangerous Harvest." Oregon's Agricultural Progress Archive. December 3, 2013. OregonProgress.OregonState.edu/winter-2003
/dangerous-harvest.

Hirst, K. Kris. "Ancient Maya Beekeeping: The Stingless Bee in Pre-Columbian America." ThoughtCo. May 11, 2018. ThoughtCo.com/ancient-maya-beekeeping-169364.

Jones, Richard, and Sharon Sweeney-Lynch. *The Beekeeper's Bible.* New York: Stewart, Tabori & Chang, 2011.

Kite, W. "The Magazine of American History with Notes and Queries 21." In *The Magazine of American History with Notes and Queries 21*, 21:523. New York: A.S. Barnes and Company, 1889.

Kritsky. Gene. *The Tears of Re: Beekeeping in Ancient Egypt.* Cambridge: Oxford University Press, 2015.

Morley, Margaret Warner. *The Honey-Makers.* Whitefish, MT: Kessinger Publishing, 1899.

Oertel, Everett. "History of Beekeeping in the United States." In *Beekeeping in the United States, Agricultural Handbook 335.* Washington DC: USDA, 1980: 2–10.

Parks, Chris. "Saint Valentine & His Beekeeping Patronage." Order of Bards, Ovates & Druids. Accessed June 5, 2020. https://druidry.org/resources/saint-valentine-his-beekeeping-patronage.

Ransome, Hilda M. *The Sacred Bee in Ancient Times and Folklore.* New York: Dover Books, 2004.

Seeley, Thomas D. *Following the Wild Bees: The Craft and Science of Bee Hunting.* Princeton, NJ: Princeton University Press, 2016.

Sullivan, Kerry. "After 2,000 Years of Harmony, the Maya May Soon Lose Their Stingless Bee Pets." Ancient Origins. June 29, 2018. Ancient-Origins .net/history-ancient-traditions/after-2000-years-harmony-maya-may -soon-lose-their-stingless-bee-pets-021360.

Synnott, Mark. "The Last Death-Defying Honey Hunter of Nepal." *National Geographic* (July 2017). NationalGeographic.com/magazine/2017/07 /honey-hunters-bees-climbing-nepal.

Virgil, Robert Wells, Walter Bachinski, and Janis Butler. *Virgil's Georgics.* Shanty Bay, ON, Canada: Shanty Bay Press, 2007.

CHAPTER 5

Lindauer. Martin. *Communication among Social Bees.* Cambridge, MA: Harvard University Press, 1961.

Seeley, Thomas D. *Honey Bee Democracy.* Princeton, NJ: Princeton University Press, 2011.

von Frisch, Karl. *The Dancing Bees: An Account of the Life and Senses of the Honey Bee* (translation of *Aus dem Leben der Bienen*), 1st English ed. New York: Harcourt Brace, 1953. Updated with new research as *The Dance Language and Orientation of Bees* (translation of *Tanzsprache und Orientierung der Bienen*), Cambridge, MA: Harvard University Press. 1967. (Also referenced in chapter 7.)

CHAPTER 6

"Bees Reared in Cities 'Healthier.'" BBC News. January 17, 2006. News.BBC .co.uk/2/hi/europe/4621184.stm.

Conrad, Ross. *Natural Beekeeping: Organic Approaches to Modern Apiculture,* 2nd ed. White River Junction, VT: Chelsea Green Publishing, 2013.

Youngsteadt, Elsa, R. Holden Appler, Margarita M. López-Uribe, David R.

Tarpy, and Steven D. Frank. "Urbanization Increases Pathogen Pressure on Feral and Managed Honey Bees." *PLoS One* 10, no. 11 (November 4, 2015): e0142031. doi.org/10.1371/journal.pone.0142031.

CHAPTER 7

Friends of the Earth. "10 Easy Ways to Help Bees in Your Garden." August 3, 2017. FriendsOfTheEarth.uk/bees/10-easy-ways-help-bees-your-garden.

Justinek, Jure. "Best Honey Plants to Help Save Bees." World Bee Day. Accessed March 10, 2020. WorldBeeDay.org/en/did-you-know/86-best-honey-plants-to-help-save-bees.html?highlight=WyJiZXN0IiwiYmVlIiwicGxhbnRzIiwicGxhbnRzJyJd.

CHAPTER 8

Fessenden, Ron, and Mike McInnes. *The Honey Revolution: Restoring the Health of Future Generations*, 2nd ed. Pearland, TX: World Class Emprise, LLC, 2009.

National Honey Board Annual Report 2012. Frederick, CO: National Honey Board, 2012. Honey.com/files/general/Annual-Report-2012_FINAL.pdf.

Traynor, Joe. *Honey: The Gourmet Medicine.* New York: Kovak Books, 2002.

Traynor, Kirsten S. *Two Million Blossoms: Discovering the Medicinal Benefits of Honey.* North Yorkshire: Image Design, 2012.

Index

Acknowledgments

Beekeeping is a journey. Mine has now extended beyond 50 years. I have been blessed to have had many teachers over the years—foremost the bees themselves and at the hands of many skilled and patient beekeepers. Some of the beekeepers that have been kind enough to mentor and share their lives with me have been real artists. My introduction to beekeeping science came courtesy of Dr. Roger Morse of Cornell University. He was kind, patient, a great mentor, and a good friend, qualities that made him a good beekeeper as well as university professor. If I have been able to pass any of those traits on to others, it is because of my beginnings with Doc.

I am still learning the science of beekeeping. I sincerely appreciate the many colleagues, students, and knowledgeable beekeepers who have made and continue to make my beekeeping journey a pleasant and fulfilling journey. Thank you all.

At times, I use language as if the bees are thinking, reasoning organisms. This is called *anthropomorphizing*—giving them human characteristics. They cannot think or reason. Although we may marvel at how they seem to be so intelligent or clever, they are merely responding to stimuli, often poorly understood by us. We have much to learn from bees. They are gentle, patient teachers, something I continue to appreciate even after 50-plus years teaching others.

Thanks to the folks at Callisto, including the very patient editor Sam Eichner.

About the Author

Dr. Dewey M. Caron is emeritus professor of entomology and wildlife ecology at the University of Delaware and affiliate professor at the department of horticulture at Oregon State University. He holds a bachelor's degree in zoology from the University of Vermont, a master's in ecology from the University of Tennessee, and a PhD in entomology (apiculture) at Cornell University. He served as a faculty member at University of Maryland from 1970 to 1981 and at the University of Delaware from 1981 to 2009, when he retired and moved to the Portland, Oregon, area to be closer to his grandkids. Dewey

continues to write for bee magazines and newsletters and gives Bee Short Courses across the country. Annually, he gives over 100 presentations to bee clubs and state bee organizations.

Dewey is also the author of *Honey Bee Biology and Beekeeping* and seven other books; he has written over 20 book chapters. He is active in master beekeeper training in Oregon and California and with the Eastern Apicultural Society on the East Coast. He represented the Western Apicultural Society on the Honey Bee Health Coalition (HBHC) and is the principle author of two reports for the HBHC, *Tools for Varroa Management* and *Best Management Practices for Bee Health.* He conducted an annual bee loss survey of the Pacific Northwest (PNW) backyard and commercial beekeepers and the annual pollination survey of PNW beekeepers.

He keeps his own backyard bee colonies in Oregon with nephew colonies of Africanized bees in Bolivia, South America, where he spends three to four months a year.

CPSIA information can be obtained
at www.ICGtesting.com
Printed in the USA
JSHW032010100622
26934JS00001B/1